INTERDISCIPLINARY MATHEMATICS

VOLUME XV

TODA LATTICES, COSYMPLECTIC MANIFOLDS, BÄCKLUND TRANSFORMATIONS AND KINKS

PART A

ROBERT HERMANN

MATH SCI PRESS
53 Jordan Road
Brookline, Ma. 02146 (USA)

ISBN 0-915692-20-1

MATH SCI PRESS
53 Jordan Road
Brookline, Ma. 02146

Printed in the United States of America

PREFACE

This volume is basically a sequel to Volumes XII and XIV, but has a more tentative research-oriented character. After developing the general viewpoint in XII and XIV, I wanted to look more carefully at the finite number-of-degrees-of-freedom models, generically called "Toda Lattices". I realized that they involved something that I know very well--"classical mechanics with constraints" on a Lie group, which is, in a sense, just the natural generalization of the rotating rigid body. A geometric structure that I had earlier (Vol. X) called a cosymplectic manifold also played a key role. Further, the basic work of Flaschka and Moser could be interpreted in a very natural way using ideas that go back to Sophus Lie--"solubility by quadratures" of a mechanical system seems to involve relating it to a solvable Lie group. In turn, the "solvable Lie groups" come in a very natural way from the Gelfand-Levitan Inverse Scattering Equation, which are "triangular" operators and hence form solvable Lie algebras. Finally, Roger Brockett noticed that Moser's solution of the Toda Lattice involved the "Impulse Response" (see Vol. VIII) of a linear, time-invariant input-output system; this led to further developments.

In short, I realized ever more forcefully that the theory of nonlinear waves and solitons is incredibly rich and ramified. Since this material seems to be of interest for current research, I decided to separate it from the more long-range material developed in Volumes XII and XIV and publish it in this way. I have also added some work I am doing on Bäcklund transformations, particularly in view towards understanding their global nature. What is the precise mechanism which enables the Bäcklund transformation to change the topological character of solutions on which it acts, particularly in view of the notion of "kinks" developed by D. Finkelstein, Perring, Rubenstein, and Skyrme? This work is the result of a long-term collaborative effort between myself, Frank Estabrook and Hugo Wahlquist (of Jet Propulsion Lab) which aimed towards a systematic understanding of the nature of a "Bäcklund transformation". I have used many of their ideas, and would like to officially thank them.

The research underlying this work was supported in part by the National Science Foundation Grant MCS-76-635.

TABLE OF CONTENTS

MY VIEWS ON APPLIED MATHEMATICS;
THE MATHEMATICIAN AS THE STRATEGIST
FOR MATHEMATICAL SCIENCE AND ENGINEERING

by Robert Hermann

Since I seem to have undergone an evolution from work in rather general and sophisticated differential geometry and topology to my present obsessive interest in the applications and ramifications of geometry in engineering and physics, I hope that my views on applied mathematics may be useful and interesting. The immediate catalyst for this essay is that I plan to lecture for several days in April at the University of North Carolina in Chapel Hill on "applied mathematics"--I hope that this document can serve as a foundation for these talks and make them available to a wider public. I have been thinking about these problems for a long time now, and I am getting sufficiently annoyed at the perversity of our establihsed elites, "pure" or "applied", to try to do something about the situation.

I have always found it easy to learn relatively abstract "modern" mathematics (say, of the Bourbaki sort) but have also had a "puritan" belief that this was not enough, that it was ultimately scientifically and intellectually sterile. One of my projects in graduate school was the study of the work of Elie Cartan on general geometric structures (G-structures, connections, infinite groups,...), which had a lasting effect on my taste and outlook. Comparing it with what was going on around me at the time in Princeton (sheaves, connections, fiber bundles, Riemann-Roch Theorems, and many other goodies--it was a Golden Age), I realized that someone like Cartan had a separate dimension that was lacking in all the "modern" stuff. Now, part of this was pure genius--Cartan had a unique ability to integrate the most detailed and concrete examples and computations with the grandiose generalization. But part was, no doubt, also the intellectual milieu in which he was educated.

What has this hero-worship to do with Applied Mathematics? Now, it is not generally known, but Cartan did in fact have a substantial background in the physics of his day. Many of his ideas have turned out to be extremely important for present-day applications. The "symplectic structure" approach to mechanics and the calculus of variations is due to him. His ideas in Relativity are just now being absorbed by the physicists--he is Godfather of the present day work on the global-singularity structure of relativity and gravitation. Above all, his work on Lie groups and algebras is fundamental for elementary particle physics. Now, one might say that this is a pure "accident"--perhaps it demonstrates the thesis that "pure mathematics" eventually will be applied. Perhaps today's papers on algebraic groups over p-adic fields will be applied to physics in 50 years, as was Cartan. Mathematicians seem to be most comfortable with the belief that the applications happen randomly, as a Poisson process; perhaps, but I don't think so! The tip-off to me is that, when Cartan heard about general relativity, he completely changed his research direction, investing ten years in the exploration of "generalized geometries". Our modern theory of connections--which is probably the only fundamental addition the 20-th century has made to differential-geometric ideas--was the result! Cartan was not at all alone among his generation of mathematicians in keeping au courant with the science of their day and jumping in where they saw something useful to do--Poincaré, Hilbert, Volterra, Levi-Civita, Caratheodory, Weyl,... Relativity and quantum mechanics were not created in a mathematical vacuum.

"Fine", you will no doubt say, "but what has this got to do with getting my student, who has written a good thesis in, say, singularities of mappings, a job teaching the FBI about catastrophes so that they can predict prison riots. That ratfink Sussman has blown the con game, and I can't even persuade my dean that I have some higher purpose in life than teaching 12 hours of calculus."

The recent essay by Sussman and Zahler (see <u>Science</u>, Vol. 196 (1977), p. 287) on the Thom-Zeeman Affair has, indeed, shown that times seem to be different; when a leading mathematician of <u>our</u> day turns his thoughts to applications, the results can be embarrassing. Perhaps these things did not happen back in Cartan's day because there were much closer links between mathematicians and the users of mathematics. Sussman and Zahler make the important point that the normal self-corrective and critical mechanism of science did not work in the case of "catastrophes" because the esoteric nature of the underlying mathematics intimidated those who were in a position to know something about the subject matter. I tend to see it as a symptom of the isolation of the world of pure mathematics--there was no intellectual continuum between the world represented by Thom and that of the biologists and social scientists--there were no competent "applied mathematicians" who were in a position by virtue of their knowledge of <u>both</u> the mathematics and the subject matter, to evaluate and develop Thom's insights. I am convinced that there <u>are</u> serious applications of the theory of singularities of mappings along the lines Thom indicates, but it will require a much more serious and detailed knowledge of the problems and methodology than Thom and Zeeman exhibit. I believe that the specific way that Thom proposes to use the mathematics is all wrong--it is much too "static" and tied in with an a-priori Keplerian mathematical metaphysics (the "harmony of the universe") that never really succeeds in science <u>in its own terms</u>. (Of course, it often inspires the Keplers to do the right thing for the wrong reasons!) Certainly, Cartan's ideas must have seemed just as mysterious to the physicists of the 1920's--in fact they have not learned them to this day--as Thom's are to today's biologist and social scientist. I can well understand the challenge that biology made to Thom and the frustration he must feel! Here was a vast area crying out for the development of new mathematical methods. The biologists were involved with the most detailed and tedious stuff that has no obvious hope of leading to an adequate general theory. Thom thought he saw a way into this general theory in terms of his own mathematics. In terms of the concept I want to develop in this essay, Thom wanted to serve as the _strategist_ of a new mathematical biology. Unfortunately for him, there were no competent _tacticians_ around.

A serious analysis of the nature (and future) of "applied mathematics" must reflect the nature of the world as it exists today. In my view, the dominant effect for the organization of applied mathematics is the existence of numbers of engineers and scientists who are reasonably well versed in the mathematics of their disciplines and have already developed a moderately sophisticated mathematical structure. One has only to pick up the IEEE journals or the Journal of Mathematical Physics to realize how things have changed in this respect in 20 years. Mathematics _is_ being applied in these areas, but it is being done in a short-term, "pragmatic" way by people who are not professional mathematicians. This has had a major effect on the employment situation for both pure and applied mathematicians. For example, my opportunity to pass over a barrier and learn something about applied mathematics came in 1959 when, after Sputnik, I was hired by an industrial laboratory because of my knowledge of nonlinear differential equations and geometry. I would not get that opportunity today--there are large numbers of well-trained engineers available, and the very nature of the work I was hired to do has changed due to the advance of computers.

However, in part due to the lack of serious involvement of professional mathematicians in large areas of applied work, I see an imbalance towards the "tactical", short-range. Now, this is also the tenor of the times--the users and financiers of mathematics (and all things for that matter) demand short-term answers and are no longer willing to support we theorists (in the style to which we became accustomed) in the hope that they would ultimately derive some benefit. The tragedy is that there is an intermediate position--instead of Radon developing his theorem in 1917, then waiting fifty five years for the engineers to realize that it is a useful tool for x-ray machines, it is possible to see mathematics develop over a five or ten year time-span to respond to the needs of science and technology. Working as far as possible in collaboration with the mathematical "tacticians" who are being trained primarily in terms of science and engineering, we can work on these intermediate range problems. For example, in systems-control theory, the electrical engineers have developed (mostly on their own) a very beautiful and useful theory of "linear systems" based on the algebra of matrices of rational functions in one variable. The classical work here is marvelously summarized in Gantmacher's "Theory of Matrices", and almost evey fact to be found there has been utilized by engineers in the recent work. There would be many uses in technology--for example in 2-D signal processing and in delay systems--for the right generalizations to matrices of rational functions in two or more variable

A systematic mathematics does not exist, but I believe it could be created--based on "modern" algebraic geometry--which is usually thought of as the most abstract and "useless" part of mathematics! Mathematical physics generates many challenges in the area of nonlinear geometry and differential equations. In my own work I am trying to think about the applications of the theory of nonlinear waves (which, thanks to the insights of Frank Estabrook and Hugo Wahlquist, two physicists who work in industry, turns out to involve Elie Cartan again!) to problems of fluid flow and fusion. Again, this will involve generalizing techniques developed for 1-D problems to two or more dimensions, where the problems become much more mathematically sophisticated and interesting.

In order to orient "applied mathematics" in this direction, I (along with Clyde Martin) proposed a new organization, the Association for Engineering and Scientific Mathematics, which would organize groups to systematically attack the mathematical problems of various areas. My own immediate interests lie with three such groups: In systems-control theory, elementary particle physics, and nonlinear waves. Since there were obvious and cogent reasons for not fragmenting the mathematical community even further, we decided to try to organize within the American Mathematical Society. This proposal is now emmeshed in their bureaucracy (I suspect it is seen as a threat by various vested interests in the AMS) and, if nothing happens soon we will carry on with our original idea of an independent organization.

We are also organizing an Institute for Engineering Scientific Mathematics which will attempt to do more long-range and directed research projects. At the moment, I am thinking of trying to organize a series of courses and symposia in the Boston area for the academic year 1977-78. My ultimate aim here is to develop short-lived research institutes which will seriously and systematically attack the mathematical problems using the full resources of modern mathematics.

I have been asked why I simply do not work within the existing structure of "applied mathematics". The sort of organized "applied mathematics" that has managed to survive in the shadow of the rather overpowering presence of pure mathematics has been, from my point of view, somewhat perverse because it has tended to mimic the structure of pure mathematics and divide up into--"techniques"--numerical analysis, combinatorics, differential equations,... just as pure mathematics divides into algebra, topology,... Now, technique _is_ important. After all, what we mathematicians have to offer--often in _competition_ with engineers and scientists--is our competence in _mathematics_. However, I think many leading applied mathematicians have an inferiority complex about the technical quality of the work of their colleagues, and have over-compensated in order to give the discipline more academic stature. In principle, I have no objection to this. My main complaint is that--in this world of diminished resources for science--they choke out the possibility of a different sort of "applied mathematics" that would be truly interactive with science and technology. For example, in one applied mathematics department in the Boston area everyone who has a serious and creative interest in interacting with other disciplines seems to be denied tenure. I have found the Applied Mathematics section of NSF to be unsympathetic to the broader view of "applied mathematics" and responsive mainly to the "technicians". I must admit that I find these "technical" applied mathematics areas very dull <u>in themselves</u>, and I feel much more at home with the attitudes of a good physicist or engineer--lets just take ideas at an intuitive level, try them out in applications, see how they work, and proceed from there. I couldn't care less about counting how many trees there are in a graph or a new algorithm for numerically solving a partial differential equation in n instead of $2n$ multiplications <u>until I see what relevance this has to the applications in which I am interested</u>. The fun for me is in developing mathematics needed directly for science, not in working in an area that someone at NYU has decided is to be called "applied mathematics" this year. I also simply get negative signals in talking to the self-defined applied mathematicians. For example, a former president of SIAM has told me that his colleagues do not regard me as a "true" applied mathematician but as someone who "teaches pure mathematics to the physicists". (In fact, my original and most important applied work--which is well known to some control engineers--was in the application of differential geometry to systems theory. And, of course, I also regard it as a higher function in life to adapt the tools of pure mathematics to the needs of the sciences!) Of course, I deeply believe in the need for a wide variety of approaches (one of my own complaints about the world of pure mathematics is its narrowness and sectarianism) and talents, but obviously if these "vested interests" continue to treat our efforts with ignorance and/or hostility, we will have no choice but to strike out on our own--I would like these facts to be known so that I am not blamed for a further dispersal of mathematics, which I recognize would be harmful to our profession as a whole.

However, these difficulties with that side are basically just misunderstandings--I believe we agree on fundamentals. The main problem is the rigidity inforced on the mathematical world by the handful of "pure" mathematics departments at our "elite" institutions. For twenty years now they have fostered a fanatical narrowness and specialization that makes today's Ph.D.'s unsuited for any other job than training more Ph.D's--that is the real reason that support for young mathematicians has so disastrously declined. As Joe Keller points out in the article from Science quoted above, those topologists who could understand Thom's topology did not have the most elementary knowledge of the state of biology or social sciences, or even of the pre-knowledge of how good science is done and tested. My own complaint is that--at the time when Lie group and geometric techniques are becoming useful across a broad spectrum of phsyics and engineering, no one among the "geometric" community (and numerically the topologists far outweigh and outvote we differential geometers) has concerned himself with the basic education of engineers and scientists in the tools of modern geometry (manifold theory, fiber bundles, etc.) nor has followed the lead of a few of us in making a serious effort to learn something about these applications. What I mainly see is a lofty superiority and condescension--an attitude that will eventually end in disaster.

CLASSICAL AND QUANTUM SOLITONS AS ELEMENTARY PARTICLES

Abstract

Classical elementary particle solitons are defined as certain "extremal" geodesics of left-invariant metrics on Lie groups. Quantum elementary particle solitons are defined in terms of the unitary representation theory of the Lie groups

1. Introduction

Ever since the development of the notion of "soliton" in the mid 1960's, there has been much speculation about their significance for elementary particle physics. Now, it seems most plausible to define "elementary particles" by group-theoretic principles. My aim here is to describe briefly a mathematical mechanism which ties together "solitons"--particularly as described by the Toda Lattice model[1,2,3]--and these group-theoretic ideas.

The immediate starting point is the observation[4] that many of the relevant physical systems could be defined in the following Lie algebraic way: Choose a Lie algebra $\mathscr{L}$, a subset $\mathscr{A} \subset \mathscr{L}$, and a map $\alpha: \mathscr{A} \to \mathscr{L}$. Look for curves $t \to a(t)$ in $\mathscr{A}$ such that

$$\frac{da}{dt} = [\alpha(a), a] \tag{1.1}$$

Now, in [5] I have also described how equations like (1.1) could be derived in terms of left invariant mechanical systems on Lie groups. (This work was a generalization of remarks by V. Arnold.[6]) In this note, I will indicate how one may describe solutions of (1.1) which seem to have the "soliton" features, using ideas of Zabusky and Kruskal.[7]

2. THE POISSON BRACKET ON THE DUAL SPACE OF A LIE ALGEBRA

I will now abstract from Réference 5 some relevant information. Suppose that the Lie algebra $\mathscr{L}$ is finite dimensional. (The ideas can be extended to the infinite dimensional case.) Suppose that the (X^i), $1 \leq i,j \leq n$ are a basis for $\mathscr{L}$, with c_k^{ij} the structure constants of $\mathscr{L}$, i.e.,

$$[X^i, X^j] = c_k^{ij} X^k$$

Introduce a linear space with coordinates (y^i), and a Poisson-bracket like operation

$$\{f_1, f_2\} = \frac{\partial f_1}{\partial y^i} \frac{\partial f_2}{\partial y^j} c_k^{ij} y^j$$

for functions on this space. It satisfies the Jacobi identity, hence defines a Lie algebra structure for the functions of y. (More invariantly, the space of y's is $\mathscr{L}$, the dual space to $\mathscr{L}^d$. This bracket derives from a "cosymplectic structure"[8] on $\mathscr{L}^d$.) If

$$\Delta = a_{i_1 \ldots i_r} X^{i_1} \ldots X^{i_r}$$

is an element of the universal enveloping algebra of $\mathscr{L}$, assign to Δ the function

$$f_\Delta = a_{i_1 \ldots i_r} y^{i_1} \ldots y^{i_r}$$

Then, we have

$$\{f_{\Delta_1}, f_{\Delta_2}\} = f_{[\Delta_1, \Delta_2]} \tag{2.1}$$

Associate to each function $f(y)$ the vector field

$$\Delta_f = \frac{\partial f}{\partial y^i} c_k^{ij} y^k \frac{\partial}{\partial y^j} \quad . \tag{2.2}$$

Then,

$$A_f(f_1) = \{f, f_1\} \tag{2.3}$$

for each function f_1 of y .

We can associate with f the orbit curves of A_f:

$$\frac{dy^k}{dt} = \frac{\partial f}{\partial y^i} c_k^{ij} y^k \tag{2.4}$$

Add to (2.4) a set of linear constraint equations:

$$y^a = 0 \quad , \qquad 1 \leq a,b \leq m \tag{2.5}$$

Assume f is quadratic in y, i.e., that it is of the following form:

$$f = b_{ij} y^i y^j \quad , \tag{2.6}$$

with constants (b_{ij}). We shall suppose that $\mathscr{A}$ and α can be chosen so that Equations (2.4)-(2.5) are equivalent to (1.1).

Now, let $f, \Delta_1, \ldots, \Delta_r$ be the Casmer operators of the Lie algebra $\mathscr{L}$, i.e., the operators of the center of the universal enveloping algebra of $\mathscr{L}$. Then, we have

$$A_f(f_{\Delta_1}) = 0 = \cdots = A_f(f_{\Delta_r}) \quad ,$$

i.e., the $f_{\Delta_1},\ldots,f_{\Delta_r}$ are constants of motion for the dynamical equations (2.4).

3. CLASSICAL SOLITONS

We are now ready to apply the definition[7] of "solitons" in terms of conserved quantities. Given a mechanical system, with Hamiltonian function f, and conserved quantities $f_1, f_2, \ldots$ which are in involution, i.e., satisfy $\{f_i, f_j\} = 0$ for $i,j = 1,2,\ldots,$ the "solitons" are the solutions of the dynamical equations determined by f which lie on the subsets obtained by extremizing (with constraints (2.5) f subject to the constraints

$$f_1 = \text{constant} \quad ,$$

then f_1, f_2 = constant, and so forth.

We can immediately carry this over to the group-theoretic situation described in Section 2, with f given by (2.6) and $f_1 = f_{\Delta_1}$, $f_2 = f_{\Delta_2}, \ldots$ Suppose that the vector field A_f given by (2.2) preserves the constraints (2.5). To obtain the "solitons", we must impose the following equations:

$$\left.\frac{\partial f}{\partial y^i} + \lambda_1 \frac{\partial f_1}{\partial y^i}\right|_{y^a=0} = 0 \quad , \qquad i = m+1,\ldots,n$$

$$\left.\frac{\partial f}{\partial y^i} + \lambda_1 \frac{\partial f_1}{\partial y^i} + \lambda_2 \frac{\partial f_2}{\partial y^i}\right|_{y^a=0} = 0 \quad , \quad i = m+1,\ldots,n$$

and so forth,

where $\lambda_1, \lambda_2, \ldots$ are Lagrange multipliers.

This type of solution has been described before[9] in a classical and quantum mechanical framework.

4. QUANTUM SOLITONS

In order to quantize this system, let L be a Lie group whose Lie algebra is $\mathscr{L}$. Treat L as the configuration space of the mechanical system. Let H be the Hilbert space of complex-valued functions on L which are square integrable with respect to the left-invariant volume element on L. Represent the y^i as left-invariant first order differential operators Y^i in H. Represent polynomials in the y's as suitably symmetrized products of the Y's. Thus, one may associate Hermitian operators $A, A_1, \ldots, A_r$ to $f, f_1, \ldots, f_r$. One may now define "quantum solitons" as those orbits of the group t exp(itA) which lie on the states which extremize

$$\frac{\langle\psi|A|\psi\rangle}{\langle\psi|\psi\rangle} ,$$

subject to the constraints $\langle\psi|A_1|\psi\rangle/\langle\psi|\psi\rangle$ = constant, $\langle\psi|A_2|\psi\rangle/\langle\psi|\psi\rangle$ = constant, etc. More details are in Ref. 9 about this procedure.

Bibliography

1. M. Toda, J. Phys. Soc. Japan, 22 (1967), p. 431

2. H. Flaschka, Phys. Rev. B9 (1974), p. 1924 and Prog. Theor. Phys. 51 (1974), p. 203.

3. J. Moser, in Lecture Notes in Physics, Vol. 38, Springer-Verlag.

4. R. Hermann, Phys. Rev. Lett. 37 (1976), p. 1591.

5. R. Hermann, J. Math. Phys. 13 (1972), p. 460.

6. V. Arnold, Ann. Inst. Grenoble, 16 (1966), p. 319.

7. N.J. Zabusky and M.D. Kruskal, Phys. Rev. Lett. 15 (1966), p. 319.

8. R. Hermann, Gauge Fields and Cartan-Ehresmann Connections, Math Sci Press, Brookline, Mass., 1975.

9. R. Hermann, J. Math. Phys. 13 (1972), p. 878.

THE DIFFERENTIAL-GEOMETRIC FOUNDATIONS OF THE INVERSE SCATTERING TECHNIQUE, I:

THE TODA LATTICE AS A LEFT-INVARIANT MECHANICAL SYSTEM ON A LIE GROUP

ABSTRACT

The equations of Toda Lattice are interpreted as generalizations of the Euler rotating rigid body equations. Both the "Lagrangian" and "Hamiltonian" structures for the general problem are described in terms of differential geometry.

1. INTRODUCTION

In [1] the following abstract pattern was established for the Korteweg-de Vries equation:

> Let $\mathscr{L}$ be a real Lie algebra, $\mathscr{A}$ a subspace of $\mathscr{L}$, $\alpha: \mathscr{A} \to \mathscr{L}$ a map. The basic dynamical equation is the following one:
>
> $$\frac{da}{dt} = [\alpha(a), a] \quad , \tag{1.1}$$
>
> to be solved for a curve $t \to a(t)$ in $\mathscr{A}$.

This equation also occurs in Euler's work in the rotating rigid body and in Arnold's work [2] on motion of perfect fluids. It is a general formulation of ideas of Lax [3], which have been fundamental for the theory of nonlinear waves and solitons. The starting point for this paper is Flaschka's [4,5] formulation of the Toda lattice (a "discrete" analogue of a nonlinear wave equation which admits the analogue of "solitons") involves an equation like (1.1), with $\mathscr{A}$ a linear subspace of $\mathscr{L}$ and α a linear map.

On the other hand, the Euler rigid body equations are basically those of a calculus of variations problem on a Lie group which is invariant under left translation [6]. As such, they have a Hamiltonian structure. One might expect that the Hamiltonian structure of the Toda lattice equations has something to do with this "natural" one; we shall, in fact, investigate their interrelation.

2. LEFT-INVARIANT CALCULUS OF VARIATIONS PROBLEMS ON A LIE GROUP

The basic references we use here are [6,7]. Let G be a connected Lie group. Let $\underset{\sim}{G}$, the Lie algebra of G, be the set of left-invariant vector fields of G. $T(G)$, the tangent vector bundle to G, can then be identified with $G \times \underset{\sim}{G}$. Namely, to each $(g,\underset{\sim}{g}) \in G \times \underset{\sim}{G}$ we associate the value

$$\underset{\sim}{g}(g) \in G_g \quad .$$

A <u>Lagrangian</u> is a map

$$L: T(G) \equiv G \times \underset{\sim}{G} \to R \quad .$$

We shall only consider those which are <u>left-invariant</u> and <u>quadratic</u> in $\underset{\sim}{G}$. To make this explicit, let

$$(\underset{\sim}{g}_i) \quad , \qquad 1 \leq i,j \leq n = \dim G$$

be a basis for $\underset{\sim}{G}$. Let

$$y^i: G \times \underset{\sim}{G} \to R$$

be the functions such that

$$y^i(g,\underset{\sim}{g})\underset{\sim}{g}^i = \underset{\sim}{g} \quad ,$$

i.e., y^i is the i-th component of $\underset{\sim}{g}$ in the basis $(\underset{\sim}{g}_i)$. Then,

$$L = \frac{1}{2} b_{ij} y^i y^j \quad , \tag{2.1}$$

where (b_{ij}) is a constant, symmetric matrix. Let θ^i be the basis of one-forms on G that is dual to the $\underset{\sim}{g}_i$, i.e.,

$$\theta^i(\underset{\sim}{g}_j) = \delta^i_j \quad . \tag{2.2}$$

They satisfy the Maurer-Cartan equations

$$d\theta^i = \frac{1}{2} c^i_{jk}\theta_j \wedge \theta_k \quad , \tag{2.3}$$

where

$$[\underset{\sim}{g}_i, \underset{\sim}{g}_j] = c^k_{ij}\underset{\sim}{g}_k \quad , \tag{2.4}$$

i.e., the (c^i_{jk}) are the structure constants of the Lie algebra $\underset{\sim}{G}$ with respect to $\underset{\sim}{g}_i$. Regard the θ^i as one-forms on $G \times \underset{\sim}{G}$, pulling them back via the projection map $G \times \underset{\sim}{G} \to \underset{\sim}{G}$. Set:

$$\omega^i = \theta^i - y^i dt \tag{2.5}$$

and

$$\omega = \frac{1}{2} b_{ij} y^i y^j dt + b_{ij} y^j \omega^i \tag{2.6}$$

ω is called the Cartan form. It lives on $G \times \underset{\sim}{G} \times T$, where T denotes the "time" manifold. The extremals of the variational problem are the Cauchy characteristic curves of $d\omega$. To determine them, proceed as follows:

$$d\omega = b_{ij} y^i dy^j \wedge dt + b_{ij} dy^j \wedge \omega^i + b_{ij} y^i d\omega^j$$

$$d\omega^j = d\theta^j - dy^j \wedge dt \quad .$$

Hence,

$$d\omega = b_{ij} dy^j \wedge \omega^i + b_{ij} y^i d\theta^j \quad .$$

Now,

$$d\theta^i = \frac{1}{2} c^i_{jk} \theta^j \wedge \theta^k$$

$$= \frac{1}{2} c^i_{jk} (\omega^j + y^j dt) \wedge (\omega^k + y^k dt)$$

$$= \frac{1}{2} c^i_{jk} \omega^j \wedge \omega^k + c^i_{jk} y^j dt \wedge \omega^k \quad .$$

Thus,

$$d\omega = (b_{kj} dy^j \wedge \omega^k + b_{ij} y^j (\tfrac{1}{2} c^i_{jk} \omega^\ell \wedge \omega^k + c^i_{\ell k} y^\ell dt \wedge \omega^k)$$

$$= \omega_k \wedge \omega^k \quad , \tag{2.7}$$

with:

$$\omega_k = b_{kj} dy^j + b_{ij} y^j c^i_{\ell k} (\tfrac{1}{2} \omega^\ell + y^\ell dt) \tag{2.8}$$

The <u>Cauchy characteristic curves</u> of d (and the extremals of the variational problem determined by L) are the curves in $G \times \underset{\sim}{G} \times T$ which are integral curves of the Pfaffian system

$$\omega^k = 0 = \omega_k \quad .$$

Thus, if $y^i(t)$ are the y-coordinates of these curves (which are parameterized by t), we have:

$$b_{kj} \frac{dy^j}{dt} + b_{ij} c^i_{\ell k} y^j y^\ell = 0 \tag{2.9}$$

To put these equations in a coordinate free form, set

$$\underset{\sim}{y}(t) = y^i(t)\underset{\sim}{g}_i \quad . \tag{2.10}$$

Define a symmetric bilinear form

$$B: \underset{\sim}{G} \times \underset{\sim}{G} \to R$$

as follows:

$$b(\underset{\sim}{g}_i, \underset{\sim}{g}_j) = b_{ij} \quad . \tag{2.11}$$

(2.9) then means that:

$$\begin{aligned} 0 &= b\left(\underset{\sim}{g}_k, \frac{d\underset{\sim}{y}}{dt}\right) + b(\underset{\sim}{g}_i, \underset{\sim}{g}_j)c^i_{\ell k}y^j y^\ell \\ &= b\left(\underset{\sim}{g}_k, \frac{d\underset{\sim}{y}}{dt}\right) + b([\underset{\sim}{g}_\ell, \underset{\sim}{g}_k], \underset{\sim}{y})y^\ell \\ &= b\left(\underset{\sim}{g}_k, \frac{d\underset{\sim}{y}}{dt}\right) - b([\underset{\sim}{g}_k, \underset{\sim}{y}], \underset{\sim}{y}) \end{aligned} \tag{2.12}$$

We can them sum up as follows:

Theorem 2.1. Let the symmetric, bilinear form $\underset{\sim}{b}$ on $\underset{\sim}{G}$ determine the left-invariant Lagrangian L, and through it the variational problem on G. Then the curves $t \to g(t)$ which are extremals of the variational problem are determined via the following differential equation:

Let $t \to \underset{\sim}{y}(t)$ be a curve in $\underset{\sim}{G}$ such that

$$b\left(\underset{\sim}{g}, \frac{d\underset{\sim}{y}}{dt}\right) = b(\underset{\sim}{y}, [\underset{\sim}{g}, \underset{\sim}{y}]) \tag{2.13}$$

for all $\underset{\sim}{g} \in \underset{\sim}{G}$.

Let $t \to g(t)$ be the curve in G such that:

$$\underset{\sim}{y}(t)(g(t)) = g'(t) \quad , \tag{2.14}$$

where $g'(t)$ denotes the tangent vector to the curve $t \to g(t)$ at t. Then, the family of curves $t \to g(t)$ determined in this way are the extremals of the variational problem.

Equation (2.13) will be called the <u>generalized Euler equation</u>. Notice that it is <u>quadratic in</u> $\underset{\sim}{y}$. This already suggests its relation to the Lax equation of inverse scattering theory. In order to see its relation more precisely, let us suppose that $\underset{\sim}{G}$ is <u>reductive</u>, i.e., the direct sum of a semisimple Lie algebra and an abelian one. In this case (and this case only) there is a <u>non-degenerate</u> symmetric bilinear form

$$B: \underset{\sim}{G} \times \underset{\sim}{G} \to R$$

which is invariant under the adjoint representation, i.e., such that

$$B([\underset{\sim}{g}_1,\underset{\sim}{g}_2],\underset{\sim}{g}_3) + B(\underset{\sim}{g}_2,[\underset{\sim}{g}_1,\underset{\sim}{g}_3]) = 0 \tag{2.15}$$

$$\text{for } \underset{\sim}{g}_1,\underset{\sim}{g}_2,\underset{\sim}{g}_3 \in G \quad .$$

Let us write b as follows:

$$b(\underset{\sim}{g}_1,\underset{\sim}{g}_2) = B(\underset{\sim}{g}_1,\beta(\underset{\sim}{g}_2)) \quad , \tag{2.16}$$

where $\beta: \underset{\sim}{G} \to \underset{\sim}{G}$ is a linear map which is symmetric with respect to the form B. Thus, we can rewrite the Euler equation (2.12) as:

$$B\left(\underset{\sim}{g}, \frac{d}{dt}\beta(\underset{\sim}{y})\right) = B(\beta\underset{\sim}{y},[\underset{\sim}{g},\underset{\sim}{y}])$$

$$= \text{, using (2.15),}$$

$$B([\underset{\sim}{y},\beta\underset{\sim}{y}],\underset{\sim}{g})$$

Thus, the Euler equations take the following form:

$$\boxed{\frac{d}{dt}\beta(\underset{\sim}{y}) = [\underset{\sim}{y},\beta\underset{\sim}{y}]} \tag{2.17}$$

which is getting closer to the Euler-Arnold-Lax equation. We can get there completely if we assume that β^{-1} exists. Set:

$$\alpha = \beta^{-1} \quad ,$$

replace $\underset{\sim}{y}$ by $\beta\underset{\sim}{y}$ in (2.17), resulting in

$$\frac{d\underset{\sim}{y}}{dt} = [\alpha\underset{\sim}{y},\underset{\sim}{y}] \quad , \tag{2.18}$$

which is, of course, precisely the Euler-Arnold-Lax eqation.

Now, the equations for the Toda lattice--as developed by Flaschka [4] and Moser [5]--involve Equation (2.17)--with $G = GL(n,R)$ --<u>together</u> with certain constraints, namely the condition that $\underset{\sim}{y}$ belong to the linear subspace of $\underset{\sim}{GL}(n,R)$ consisting of the Jacobi matrices, i.e., the tridiagonal matrices which are symmetric. We shall now investigate this phenomenon in its general setting.

3. THE EULER EQUATIONS WITH CONSTRAINTS

Continue with the general setting described in Section 2. $\underset{\sim}{G}$ is a real, finite dimensional Lie algebra, with a non-degenerate adjoint-invariant symmetric bilinear form $B: \underset{\sim}{G} \times \underset{\sim}{G} \to R$. A is a linear subspace of $\underset{\sim}{G}$, $\alpha: \underset{\sim}{A} \to \underset{\sim}{G}$ a linear map, and the equations

$$\frac{d\underset{\sim}{a}}{dt} = [\alpha(\underset{\sim}{a}), \underset{\sim}{a}] \tag{3.1}$$

are given to us.

Now, if α can be extended as a <u>symmetric</u> (with respect to B) linear map $\alpha': \underset{\sim}{G} \to \underset{\sim}{G}$, then the equations (3.1) form a system of equations of Euler's type, namely:

$$\frac{d\underset{\sim}{g}}{dt} = [\alpha'(\underset{\sim}{g}), \underset{\sim}{g}] \quad . \tag{3.2}$$

Precisely, the solutions of (3.1) are the set of solutions $t \to \underset{\sim}{g}(t)$ of (3.2) such that $\underset{\sim}{g}(t) \in \underset{\sim}{A}$ for all t. In turn, we know that (3.2) are obtained as extremal equations of a quadratic, left-invariant variational problem on G. Here is one condition that enables α to be extended in this way.

<u>Theorem 3.1</u>. Let $\underset{\sim}{A}$ be a linear subspace of $\underset{\sim}{G}$ such that the form B is non-degenerate when restricted to $\underset{\sim}{A}$. Let

$\underset{\sim}{A}^{\perp} = \{\underset{\sim}{g} \in \underset{\sim}{G}: B(\underset{\sim}{g},\underset{\sim}{A}) = 0\}$, the <u>orthogonal complement</u> of $\underset{\sim}{A}$ in $\underset{\sim}{G}$ with respect to $\underset{\sim}{G}$. Then, B is non-degenerate when restricted to $\underset{\sim}{A}^{\perp}$, and $\underset{\sim}{G}$ is the direct sum of $\underset{\sim}{A}$ and $\underset{\sim}{A}^{\perp}$. Further, let

$$\alpha: \underset{\sim}{A} \to \underset{\sim}{A}^{\perp}$$

be a linear map. Then, it can be extended to a linear map $\alpha': \underset{\sim}{G} \to \underset{\sim}{G}$ which is symmetric with respect to the form B.

<u>Proof</u>. The first statement about $\underset{\sim}{A}$ and $\underset{\sim}{A}^{\perp}$ are well known facts of linear algebra.

To prove the extendability of α, let $\alpha^T: \underset{\sim}{A}^{\perp} \to \underset{\sim}{A}$ be the transpose of α with respect to B, i.e.,

$$B(\alpha^T(\underset{\sim}{a}^{\perp}),\underset{\sim}{a}) = aB(a^{\perp},\alpha^T(\underset{\sim}{a})) ,$$

$$\text{for } \underset{\sim}{a} \in \underset{\sim}{A}; \quad \underset{\sim}{a}^{\perp} \in \underset{\sim}{A}^{\perp} .$$

Now set:

$$\alpha'(\underset{\sim}{a}) = \alpha(\underset{\sim}{a}) , \qquad \text{for } \underset{\sim}{a} \in \underset{\sim}{A} ,$$

$$\alpha'(\underset{\sim}{a}^{\perp}) = \alpha^T(\underset{\sim}{a}^{\perp}) , \qquad \text{for } \underset{\sim}{a}^{\perp} \in \underset{\sim}{A}^{\perp} .$$

Finally, we must show that α' is symmetric with respect to B. The following identities do the job:

$$B(\alpha'(\underset{\sim}{a}),\underset{\sim}{a}_1) = B(\alpha(\underset{\sim}{a}),\underset{\sim}{a}_1) = 0 = B(\underset{\sim}{a},\alpha'(\underset{\sim}{a}_1))$$

$$B(\alpha'(\underset{\sim}{a}),\underset{\sim}{a}_1^{\perp}) = B(\alpha(\underset{\sim}{a}),\underset{\sim}{a}_1^{\perp}) = B(\underset{\sim}{a},\alpha^T(\underset{\sim}{a}_1^{\perp})) = B(a,\alpha'(a_1^{\perp}))$$

$$B(\alpha'(\underset{\sim}{a}^{\perp}),\underset{\sim}{a}_1^{\perp}) = B(\alpha^T(\underset{\sim}{a}^{\perp}),\underset{\sim}{a}_1^{\perp}) = 0 = B(\underset{\sim}{a}^{\perp},\alpha'(\underset{\sim}{a}_1^{\perp})) \quad .$$

$$B(\alpha'(\underset{\sim}{a}^{\perp}),\underset{\sim}{a}_1) = B(\alpha^T(\underset{\sim}{a}^{\perp}),\underset{\sim}{a}_1) = B(\underset{\sim}{a}^{\perp},\alpha(\underset{\sim}{a}_1))$$

$$= B(\underset{\sim}{a}^{\perp},\alpha'(\underset{\sim}{a}_1)) \quad .$$

We can apply this construction to the Toda Lattice equations [4,5]. Let $\underset{\sim}{G}$ = the Lie algebra of $n \times n$ real matrices, $\underset{\sim}{A}$ = linear subspace of Jacobi matrices, i.e., those of the symmetric, tri-diagonal form, i.e., the form:

$$M = \begin{pmatrix} b_1 & a_1 & 0 & \cdots & & \\ a_1 & b_2 & a_2 & 0 & \cdots & \\ 0 & a_2 & b_3 & a_3 & 0 & \cdots \\ \vdots & \vdots & & & & \\ & & & & b_n & \end{pmatrix}$$

Let $\alpha\colon \underset{\sim}{A} \to \underset{\sim}{G}$ map this matrix into the following tri-diagonal skew-symmetric matrix:

$$\alpha(M) = \begin{pmatrix} 0 & a_1 & \cdots \\ -a_1 & 0 & \cdots \\ \vdots & & 0 \end{pmatrix} \tag{3.3}$$

The inner product B for $\underset{\sim}{G}$ is the following one:

$$B(M_1,M_2) = \text{trace}(M_1 M_2)$$

It is readily seen that, in this metric, the symmetric matrices are orthogonal to the skew-symmetric ones. Hence,

$$\alpha(\underset{\sim}{A}) \subset \underset{\sim}{A}^{\perp} \quad .$$

Theorem 3.1 applies, and α can be extended to a map α'. Notice that the α's are not one-one. This implies (as we shall see in Section 4) that the corresponding variational problem on G is "singular". We can then sum up as follows.

<u>Theorem 3.2</u>. Let $\underset{\sim}{A}$ be the set of real $n \times n$ Jacobi matrices. Let $\alpha\colon \underset{\sim}{A} \to \underset{\sim}{G}$ the map defined by formula (3.3), so that

$$\frac{dM}{dt} = [\alpha(M),M] \tag{3.4}$$

are the dynamical equations for the Toda Lattices as developed by Flaschka [4]. Then, there is an extension of α to a symmetric map $\alpha'\colon \underset{\sim}{G} \to \underset{\sim}{G}$ such that the set of solutions of (3.4) is identified with the set of solutions of

$$\frac{dM}{dt} = [\alpha'(M)M] \tag{3.5}$$

such that $M(t) \in \underset{\sim}{A}$ for all t.

4. LEFT INVARIANT HAMILTONIAN SYSTEMS ON LIE GROUPS

Let us return to a general Lie algebra $\underset{\sim}{G}$ which has a symmetric, non-degenerate, adjoint-invariant symmetric bilinear form $B: \underset{\sim}{G} \times \underset{\sim}{G} \to R$. If $\alpha: \underset{\sim}{G} \to \underset{\sim}{G}$ is symmetric with respect to B, one forms the equations:

$$\frac{d\underset{\sim}{g}}{dt} = [\alpha(\underset{\sim}{g}), \underset{\sim}{g}] \tag{4.1}$$

Now, if $\beta = \alpha^{-1}$ exists, we know (Section 2) how to interpret the solutions of (4.1) as extremals of a left-invariant variational problem on G. For a general, non-invertible α (such as one encounters in studying the Toda Lattice equations, as explained in Section 3), one can interpret the variational problem as a "singular" one. (For example, for the rotating rigid body this corresponds to one of the moments of inertia going to infinity.) However, there is an alternative (and dual) "Hamiltonian" interpretation. It is this to which we turn. (The notation will be slightly different from those used previously for the "Lagrangian" approach.)

The cotangent bundle to the Lie group G is just

$$G \times \underset{\sim}{G}^d \quad ,$$

where $\underset{\sim}{G}^d$ denotes the dual space to $\underset{\sim}{G}$. Let us describe a basis for forms, and write the natural contact structure in terms of this basis.

Let $(\underset{\sim}{g}_i)$, $1 \leq i,j \leq n$, be a basis for $\underset{\sim}{G}$. Let $(\underset{\sim}{g}^i)$ be the dual basis for $\underset{\sim}{G}^d$, i.e.,

$$\underset{\sim}{g}^i(\underset{\sim}{g}_j) = \delta^i_j \quad .$$

This basis $(\underset{\sim}{g}^i)$ defines linear functions which we label as z_i on $\underset{\sim}{G}^d$. Pull them back (via the Cartesian projection map $G \times \underset{\sim}{G}^d \to \underset{\sim}{G}^d$) to be functions--which we also label as z_i--on $G \times \underset{\sim}{G}^d$. Similarly, the $\underset{\sim}{g}^i$ define left-invariant one-forms on G, that we label as θ^i, and pull-back to consider as one-forms on $G \times \underset{\sim}{G}^d$. Set:

$$\theta = z_i \theta^i \quad . \tag{4.2}$$

It is the <u>contact one-form</u> on $G \times \underset{\sim}{G}^d$. Note that:

$$d\theta^i = \frac{1}{2} c^i_{jk} \theta^j \wedge \theta^k \quad , \tag{4.3}$$

where (c^i_{jk}) are the <u>structure constants</u> of the Lie algebra $\underset{\sim}{G}$. Thus,

$$d\theta = dz_i \wedge \theta^i + \frac{1}{2} z_i c^i_{jk} \theta^j \wedge \theta^k \tag{4.4}$$

If f is a smooth real-valued function on $G \times \underset{\sim}{G}^d$, let A_f be the vector field such that:

$$df = A_f \lrcorner \, d\theta \ . \tag{4.5}$$

The orbit curves of A_f are the solutions of the <u>Hamilton equations with</u> f <u>as Hamiltonian</u>. The Poisson bracket of two such functions is then defined as

$$\{f,g\} = -A_f(g)$$

In particular, suppose that f is given by a quadratic form on $\underset{\sim}{G}^d$, i.e., it is of the form

$$f = \frac{1}{2} b^{ij} z_i z_j \tag{4.6}$$

$$df = b^{ij} z_i dz_i$$

$$= A_f \lrcorner \, d\theta$$

$$= A_f(z_i)\theta^i - \theta^i(A_f)dz_i + z_i c^i_{jk}\theta^j(A_f)\theta^k$$

Hence:

$$\theta^i(A_f) = -b^{ij}z_j \tag{4.7}$$

$$A_f(z_i) = -z_k c^k_{ji}\theta^j(A_f)$$

$$= z_k c^k_{ji} b^{j\ell} z_\ell$$

The differential equations for the orbit curves of A_f are then:

$$\frac{dz_i}{dt} = z_k z_\ell c^k_{ji} b^{j\ell} \tag{4.8}$$

They are also of the "Euler" type. In order to put them there exactly, we form B on $\underset{\sim}{G}$ to dualize.

Let us put Equations (4.8) into coordinate-free terms. Recall that $(\underset{\sim}{g}_i)$ is the basis for $\underset{\sim}{G}$, $(\underset{\sim}{g}^i)$ is the dual basis for $\underset{\sim}{G}^d$. Let

$$\underset{\sim}{b}: \underset{\sim}{G}^d \to \underset{\sim}{G}$$

be the linear map defined by the following formula:

$$\underset{\sim}{b}(\underset{\sim}{g}^i) = b^{ij}\underset{\sim}{g}_j \quad . \tag{4.9}$$

Set:

$$\underset{\sim}{z}(t) = z_i(t)\underset{\sim}{g}^i \in \underset{\sim}{G}^d \qquad (4.10)$$

Thus,

$$\frac{d\underset{\sim}{z}}{dt} = \frac{dz_i}{dt}\,\underset{\sim}{g}^i$$

$$= \text{, using (4.8),}$$

$$z_k z_\ell c^k_{ji} b^{j\ell} \underset{\sim}{g}^i$$

$$= z_k z_\ell \underset{\sim}{g}^k([\underset{\sim}{g}_j, \underset{\sim}{g}_i]) b^{j\ell} \underset{\sim}{g}^i$$

$$= \underset{\sim}{z}([z_\ell \underset{\sim}{b}(\underset{\sim}{g}\ell), \underset{\sim}{g}_i])\underset{\sim}{g}^i$$

$$= \underset{\sim}{z}([b(\underset{\sim}{z}), \underset{\sim}{g}_i])\underset{\sim}{g}^i \qquad (4.11)$$

This is another, more coordinate-free version of the Euler equations. Now, work with the non-degenerate, symmetric bilinear form

$$B\colon \underset{\sim}{G} \times \underset{\sim}{G} \to R$$

which is invariant under the adjoint representation. Let

$$B_{ij} = B(\underset{\sim}{g}_i, \underset{\sim}{g}_j) \quad . \qquad (4.12)$$

Let $\underset{\sim}{B}\colon \underset{\sim}{G}^d \to \underset{\sim}{G}$ be the linear map defined by the formula:

$$\underset{\sim}{B}(g^i) = B^{ij} g_j \quad , \qquad (4.13)$$

where (B^{ij}) is the inverse matrix to (B_{ij}). Apply $\underset{\sim}{B}$ to both sides of (4.11):

$$\begin{aligned}
\underset{\sim}{B}\,\frac{d\underset{\sim}{z}}{dt} &= \frac{d}{dt}\,\underset{\sim}{B}(\underset{\sim}{z}) \\
&= \underset{\sim}{z}([\underset{\sim}{b}(\underset{\sim}{z}),\underset{\sim}{g}_i])\underset{\sim}{B}(\underset{\sim}{g}^i) \\
&= B(\underset{\sim}{B}(\underset{\sim}{z}),[\underset{\sim}{b}(\underset{\sim}{z}),\underset{\sim}{g}_i])\underset{\sim}{B}(\underset{\sim}{g}^i) \\
&= -B([\underset{\sim}{b}(z),B(\underset{\sim}{z})]g_i)\underset{\sim}{B}(g^i) \\
&= B([\underset{\sim}{B}(\underset{\sim}{z}),\underset{\sim}{b}\underset{\sim}{B}^{-1}(\underset{\sim}{B}(\underset{\sim}{z}))],\underset{\sim}{g}_i)B(\underset{\sim}{g}^i)
\end{aligned}$$

Set:

$$\underset{\sim}{y}(t) = \underset{\sim}{B}\underset{\sim}{z}(t) \tag{4.14}$$

Then, we see that it satisfies the following differential equation:

$$\frac{d\underset{\sim}{y}}{dt} = [\underset{\sim}{y},\alpha\underset{\sim}{y}] \quad , \tag{4.15}$$

with:

$$\alpha = \underset{\sim}{b}\underset{\sim}{B}^{-1} \tag{4.16}$$

This is, indeed, precisely the Lax equation of Inverse Scattering Theory applied to the case

$G = GL(n,R)$,

α = map constructed from the map: (Jacobi matrices) → (skew-symmetric, tridiagonal matrices) indicated above.

These are precisely the Toda Lattice equations, when supplemented by the additional condition that:

$$\underset{\sim}{y}(t) \in \text{Jacobi matrices for all } t \quad .$$

The next step on the agenda--in Part II of this series--is to study what this form of the equations implies about the existence of <u>integrals</u> of <u>motion</u>. In fact, we shall see that just as for the Rotating Rigid Body, where there is one constant of motion in addition to the total energy, namely, total angular momentum, which is also the Casimer operator of $\underbrace{SO(3,R)}$, that the center of the Universal Enveloping Algebra of $\underset{\sim}{G}$ provides constants of motion of the Equations (4.15). It will follow <u>automatically</u> from the formalism that these constants of motion are "in involution" with respect to the natural symplectic structure. This will provide for the "solubility" in the classical sense [8] of the equations (what other authors call--wrongly--"complete integrability".)

BIBLIOGRAPHY

1. R. Hermann, Phys. Rev. Lett. 37 (1976), p. 1591.

2. V. Arnold, Sur la géométrie différentielle des groupes de Lie de dimension infinie et ses applications à l'hydrodynamique des fluides parfaits, Ann. Inst. Grenoble, 16, 319-361 (1966).

3. P. Lax, Integrals of nonlinear equations of evolution and solitary waves, Comm. Pure Appl. Math. 21 (1968) 467-490.

4. H. Flaschka, On the Toda Lattice, I & II, Phys. Rev. B9 (1974), 1924-1925, and Prog. Theor. Phys. 51 (1974), 703-716.

5. J. Moser, in Lecture Notes in Physics, Vol. 38, ed., J. Moser, Springer-Verlag, 1975.

6. R. Hermann, Geodesics and classical mechanics on Lie groups, J. Math. Phys, 13 (1972), p. 460.

7. R. Hermann, Differential Geometry and the Calculus of Variations, Academic Press, 1968.

8. E.T. Whittaker, Analytic Dynamics, Cambridge University Press, London, 1959.

THE DIFFERENTIAL-GEOMETRIC FOUNDATIONS OF THE INVERSE SCATTERING TECHNIQUE, II:

COSYMPLECTIC STRUCTURES ON LIE ALGEBRAS AND "FUNCTION GROUPS" IN THE SENSE OF SOPHUS LIE

Abstract

Some general features of the differential geometry of cosymplectic manifolds which are useful for the study of Toda lattice equations and their generalizations are presented.

1. INTRODUCTION

In Part I, I have put the "Toda lattice" into general Lie algebra setting:

$$\frac{d\underset{\sim}{y}}{dt} = [\underset{\sim}{y}, \alpha(\underset{\sim}{y})] \quad , \tag{1.1}$$

where $t \to \underset{\sim}{y}(t)$ is a curve in a Lie algebra $\underset{\sim}{G}$, $\alpha\colon \underset{\sim}{G} \to \underset{\sim}{G}$ is a linear map. This is an equation that has appeared in the work of Euler (i.e., the rotating rigid body), Arnold [1] and Lax [2]. I have also discussed conditions that this system have a natural Hamiltonian structure associated to it. When this occurs, we of course have available all of the machinery of classical mechanics, e.g., the relation between "symmetries" and "conservation laws". In particular, one might ask for functions covered under the flow generated by (1.1) <u>which are functions of $\underset{\sim}{y}$ above</u>. It turns out that there is a special feature available here--the Poisson bracket (with respect to the underlying symplectic structure) of two such functions of $\underset{\sim}{y}$ is a function of $\underset{\sim}{y}$ again. This is a reflection of a general property of Lie algebra I have described in [3]--the <u>dual space of a Lie algebra has a</u> "<u>cosymplectic structure</u>", i.e., a skew-symmetric, <u>two-contravariant</u> tensor field which is, in a certain sense, "integrable". (Recall that a "symplectic structure" is a skew-symmetric, two <u>covariant</u> tensor field, i.e., a two-form, which is "integrable" in the sense that its exterior derivative is zero.) In this Part II I shall briefly develop the general theory of such cosymplectic structures, describing the relation to what Sophus Lie called "function groups" (in fact, it is more-or-less identical!), then proceed to the equations of form (1.1).

2. COSYMPLECTIC STRUCTURES AND THEIR HOMOMORPHISMS

Let M be a manifold. $F(M)$ denotes its C^∞, real-valued functions. $F^1(M)$ denotes its one-differential form, considered as an $F(M)$-module. Consider a two-contravariant, skew-symmetric tensor field on M, i.e., a skew-symmetric, $F(M)$-bilinear map

$$\omega\colon F^1(M) \times F^1(M) \to F(M)$$

It defines a map

$$(f_1, f_2) \to \{f_1, f_2\} = \omega(df_1, df_2) \tag{2.1}$$

of $F(M) \times F(M) \to F(M)$. ω is said to define a <u>cosymplectic structure</u> if this map $F(M) \times F(M) \to F(M)$ satisfies the Jacobi identity, i.e., defines $F(M)$ as a Lie algebra. In this case, it is called the <u>Poisson bracket</u> defined by the cosymplectic structure.

Suppose such a cosymplectic form ω is given. It is said to be <u>non-degenerate</u> if the following condition is satisfied:

$$\omega(\theta, F^1(M)) = 0 \Rightarrow \theta = 0$$

In this case, ω can be "dualized" to define a closed two-differential form on M, i.e., a symplectic structure in the usual sense.

Whenever one studies a general "mathematical structure", one should also--following the insight of the "category" theorists--study their "morphisms". Surprisingly, this insight has not been followed up as systematically in differential geometry as it might have been. (For example, the full concept of "homomorphism" of a Riemannian metric was introduced in [4]; Rienhart's notion of "bundle-like metric" is a specialization. What we are doing here is essentially just the "symplectic" version of this.) Here is the relevant concept for "cosymplectic structures".

Let M,M' be manifolds, with cosymplectic structures ω and ω'. A map $\alpha\colon M \to M'$ is said to be a <u>homomorphism</u> between the symplectic structures if the following condition is satisfied:

$$\phi^*(\omega'(\theta_1',\theta_2')) = \omega(\phi^*(\theta_1'),\phi^*(\theta_2')) \tag{2.2}$$

$$\text{for } \theta_1',\theta_2' \in F^1(M) \quad .$$

In [3] this notion is used--when specialized--to show the relation between the cosymplectic structure on the dual space to a Lie algebra and the <u>symplectic</u> structure on the dual adjoint orbits whose theory was most fully developed by Kostant [5]. ("Homogeneous symplectic structures" had been extensively studied in the 1950's, particularly by A. Lichnerowicz

Kostant's contribution was to fully appreciate how they could be embedded in the dual space to the Lie algebra, and to develop certain relations between this "picture" and Lie group representation theory--quantum mechanics. However, it was a bit of an oversell to call the business "quantization", and perhaps harmful to overall intellectual interaction between mathematicians and physicists, since what Kostant means by "quantization" is not necessarily what the physicists (or Nature) means!)

3. THE DYNAMICS ASSOCIATED WITH A COSYMPLECTIC STRUCTURE

Let $M,\omega\colon F^1(M) \times F^1(M) \to F(M)$ continue as before. For $f \in F(M)$, let A_f be the vector field on M defined by the following formula:

$$A_f(f') = \omega(df,df') \equiv \{f,f'\} \tag{3.1}$$

for $f' \in F(M)$.

Consider the orbit curves of A_f, i.e., the curves $t \to p(t)$ in M such that

$$\frac{dP}{dt} = A_f(p(t)) \quad . \tag{3.2}$$

They satisfy the following "Liouville" equation:

$$\frac{d}{dt} f'(p(t)) = \{f,f'\}(p(t)) \tag{3.3}$$

Formula (3.3) is the typical relation between "conservation laws" and "symmetries".

4. THE COSYMPLECTIC STRUCTURE ON THE DUAL SPACE TO A REAL LIE ALGEBRA

Let $\mathscr{L}$ be a real (finite dimensional) Lie algebra. In particular, $\mathscr{L}$ is, of course, a real vector space. Let $\mathscr{L}^d$ be the dual space to this vector space. Since it too is a finite dimensional real vector space, it is a manifold. For each $p \in \mathscr{L}^d$, tje tangent space $\mathscr{L}^d_p$ is identified with $\mathscr{L}^d$ itself. Hence, the space of one-covectors to $\mathscr{L}^d$ at p is equal to $\mathscr{L}$. Thus, if $\theta \in F^1(\mathscr{L}^d)$, its value $\theta(p)$ at p can be identified with an element of $\mathscr{L}$, which is, of course, given to us as a Lie algebra. Define:

$$\omega(\theta_1,\theta_2)(p) = p([\theta_1(p),\theta_2(p)]) \tag{4.1}$$

It is shown in [3] that this defines cosymplectic structures for $\mathscr{L}^d$. Here is its relation to the Lie algebra structure with which we began.

Let $U(\mathscr{L})$ denote the universal envelope associative algebra to $\mathscr{L}$. Recall that $U(\mathscr{L})$ is the quotient of the tensor algebra $T(\mathscr{L})$ by the two-sided ideal generated by all elements of the form:

$$\ell_1 \otimes \ell_2 - \ell_2 \otimes \ell_1 - [\ell_1,\ell_2] \tag{4.2}$$

$U(\mathscr{L})$ has an ascending filtration

$$R \equiv U^0(\mathscr{L}) \subset U^1(\mathscr{L}) \equiv \mathscr{L} \subset U^2(\mathscr{L}) \subset \cdots$$

namely, one says that an element has filtration r if it can be written in at least one way as a polynomial of degree at most r in the elements of $\mathscr{L}$. $U_{gr}(\mathscr{L})$ denotes the graded object, i.e.,

$$U^r_{gr}(\mathscr{L}) = U^r(\mathscr{L})/U^{r-1}(\mathscr{L}) \quad .$$

The associative algebra structure on $U(\mathscr{L})$ passes to the quotient to define a <u>commutative associative</u> algebra structure on $\mathscr{L}$. (This is essentially just the symmetric algebra $S(\mathscr{L})$ of the vector space $\mathscr{L}$.) However, the Lie algebra structure on $U(\mathscr{L})$ also passes to the quotient to define a Lie algebra structure on $U_{gr}(\mathscr{L})$. We shall next see that this Lie algebra structure is closely related to the cosymplectic structure on $\mathscr{L}^d$.

Associate with each homogeneous element $\Delta = \ell_1 \otimes \cdots \otimes \ell_r$ of $T(\mathscr{L})$ the function

$$f_\Delta \colon \mathscr{L}^d \to R$$

by the following formula:

$$f_\Delta(p) = p(\ell_1)\cdots p(\ell_2) \qquad (4.3)$$

for $p \in \mathscr{L}^d$.

It is readily seen that the map

$$\Delta \to f_\Delta$$

passes to the quotient to define a map

$$U_{gr}(\mathscr{L}) \to F(\mathscr{L}^d)$$

such that:

$$f_{\Delta_1} f_{\Delta_2} = f_{\Delta_1 \Delta_2} \tag{4.4}$$

$$\{f_{\Delta_1}, f_{\Delta_2}\} = f_{[\Delta_1, \Delta_2]} \tag{4.5}$$

The product $\Delta_1 \Delta_2$ on the right hand side of (4.4) is that of the commutative multiplication on $U_{gr}(\mathscr{L})$. The bracket $\{\ \}$ in (4.5) is the Poisson bracket defined by the cosymplectic structure on $\mathscr{L}^d$. $[\Delta_1, \Delta_2]$ on the right hand side of (4.5) is the Lie algebra bracket on $U_{gr}(\mathscr{L})$. We can restate this as follows:

> $U_{gr}(\mathscr{L})$, with its two algebra structures, is isomorphic to the polynomial functions: $\mathscr{L}^d \to R$, under point-wise multiplication and Poisson bracket.

5. THE CASIMIR OPERATORS OF A LIE ALGEBRA AS CONSTANTS OF MOTION

Continue with $\mathscr{L}$ as before. Let

$$PF(\mathscr{L}^d)$$

be the maps: $\mathscr{L}^d \to R$ which are polynomials with respect to the vector space structure. What we have essentially shown is that the Poisson bracket operator defined by the cosymplectic structure on $\mathscr{L}^d$ satisfies:

$$\{PF(\mathscr{L}^d), PF(\mathscr{L}^d)\} \subset PF(\mathscr{L}^d) \tag{5.1}$$

In particular, the polynomial functions--which are isomorphic to $S(\mathscr{L})$, the symmetric tensor algebra on $\mathscr{L}$--admit a Lie algebra structure. This Lie algebra is just, of course, the Lie algebra structure defined in $U_{gr}(\mathscr{L})$, the graded universal envoloping algebra.

Let $h \in PF(\mathscr{L}^d)$. Define a vector field A_h on $\mathscr{L}^d$ by the formula:

$$A_h(f) = \{h,f\} \quad .$$

Consider the orbit curves

$$\frac{d\underset{\sim}{y}}{dt} = A_h(\underset{\sim}{y}) \tag{5.2}$$

of h. We have seen that by choosing h to be a quadratic function, Equations (5.1) can have the form of the Toda lattice equations.

An element $f \in PF(\mathscr{L}^d)$ is a <u>Casimir-operator</u> of $\mathscr{L}$ if it lies on the center of $PF(\mathscr{L}^d)$, i.e.,

$$\{f, PF(\mathscr{L}^d)\} = 0 \tag{5.3}$$

In particular, we have:

$$\frac{d}{dt} f(\underset{\sim}{y}(t)) = 0 \tag{5.4}$$

for all solutions $t \to \underset{\sim}{y}(t)$ of (5.2), i.e., f is a constant of motion for the Equations (5.2).

If two such Casimir constants of motion f_1 and f_2 are given, then

$$\{f_1, f_2\} = 0 \quad ,$$

i.e., they are "in involution" in the sense of classical mechanics. Let

$$f_1, \ldots, f_2$$

be a maximal set of such functionally independent Casimir elements. (If $\mathscr{L}$ is semisimple, then r = rank of $\mathscr{L}$.) If $n = \dim \mathscr{L}$, and if

$$(n-1) = 2r \quad , \tag{5.5}$$

then the system (5.4) is "soluble" in the classical sense [6], with the f's providing the "action variables". For semi-simple algebras, (5.5) is only satisfied for limited cases of n, since typically n is of the order of r^2. For example, for the rigid rotating body, $\mathscr{L} = SO(3,\underset{\smile}{R})$, $n = 3$, $r = 1$.

The "Toda lattice" [7,8,9] is "soluble" in a more general sense. Namely, there is a submanifold

$$N \subset \mathscr{L}^d \quad ,$$

with $\dim N = n$, such that (5.5) is satisfied, and such that:

$$A_h \ \underline{\text{is tangent to}}\ N \quad . \tag{5.6}$$

(In the Toda lattice case, $\mathscr{L} = \underset{\sim}{GL(r,R)}$, $\mathscr{L}$ is identified with $\mathscr{L}^d$ using the trace form, and N is the linear subspace of the Jacobi matrices, i.e., the symmetric tridiagonal matrices.) The f's are the symmetric functions of the eigenvalues of the matrices. Thus, we have absolutely "free" the property that plays such a basic role in the work of Flaschka and Newell, namely, that the eigenvalues have zero Poisson brackets with each other.

An obvious basic problem is to:

Classify the linear subspaces $N \subset \mathscr{L}^d$ satisfying these conditions.	(5.7)

The article by Moser [9] seems to give several examples, closely linked to systems which admit solitons.

Lest the reader think that all of these beautiful ideas have been the result of the unique brilliance of present-day mathematics, I now want to point out that Sophus Lie would have understood all of this very well--in fact, in his prime he probably would have tossed it off in an afternoon if he had been so motivated. There are intimate relations to objects he called "function groups", which are not "groups" in the modern sense, but Lie algebras.

6. SOPHUS LIE'S "FUNCTION GROUPS" AND COSYMPLECTIC MANIFOLDS

Most of Lie's work in this area is collected together in Volume 2 of "Transformationsgruppen" [10]. Here is a translation into modern notation.

Let M be a symplectic manifold, i.e., M is even dimensional and carries a closed two-form ω which has no nonzero Cauchy characteristic vectors. $F(M)$ denotes the C^∞, real-valued functions on M. The two-form ω defines a Poisson bracket structure on $F(M)$. Here is one way it is defined.

Since ω is non-degenerate, it defines an $F(M)$-linear isomorphism

$$\eta\colon V(M) \to F^1(M)$$

($V(M)$ denotes the vector fields on M; $F^1(M)$ denotes the one-forms.) η is related to ω via the following formula:

$$\omega(v_1,v_2) = \eta(v_1)(v_2) \quad \text{for } v_1,v_2 \in V(M) \quad .$$

Let $\omega'\colon F^1(M) \times F^1(M) \to F(M)$ be the cosymplectic structure such that:

$$\omega(v_1,v_2) = \omega'(\eta(v_1), \eta(v_2)) \tag{6.1}$$

Then, for $f_1,f_2 \in F(M)$,

$$\{f_1,f_2\} = \omega'(df_1,df_2) \quad . \tag{6.2}$$

Now, a <u>function group</u> in the sense of Lie is a linear subspace $\mathscr{F} \subset F(M)$ with the following properties:

a) $\{\mathscr{F}, \mathscr{F}\} \subset \mathscr{F}$, i.e., $\mathscr{F}$ is a Lie subalgebra of the Poisson bracket structure on $F(M)$.

b) If $f_1, \ldots, f_n \in \mathscr{F}$, and if $F: R^n \to R$ is a C^∞ map, then the composite function

$$F(f_1, \ldots, f_n)$$

also belongs to $\mathscr{F}$.

Here is one way to generate such function groups. Let M'' be another manifold, and let

$$\phi: M \to M''$$

be a submersion map, i.e., a smooth map whose differential $\phi_*: T(M) \to T(M'')$ is onto. Set:

$$\mathscr{F} = \phi^*(F(M''))$$

We must now examine the conditions that ϕ must satisfy in order that $\mathscr{F}$ define a "function group", in Lie's sense. Suppose that M'' has a cosymplectic structure defined by a form $\omega'': F^1(M'') \times F^1(M'') \to F(M'')$. Suppose also that ϕ defines a homomorphism between the cosymplectic structures defined by ω' and ω''. Then, it is readily seen that $\mathscr{F} = \phi^*(F(M''))$ is a function group. Conversely, it is readily seen that if $\mathscr{F}$

is a "function group" then there is a cosymplectic structure on M" with these properties. (For example, this would proceed using the technique described in Chapter 1 of Geometry, Physics and Systems [11]. The "function group" property implies that there is a "Poisson bracket" structure on F(M"). Its "symbol", as a differential operator, then would define the "cosymplectic structure" on M".

7. THE COSYMPLECTIC STRUCTURE ON THE DUAL OF A LIE ALGEBRA AS AN EXAMPLE OF A "FUNCTION GROUP"

It is well known that if M is a manifold, its cotangent bundle $T^d(M)$ has a symplectic structure. This plays a basic role in classical mechanics. One could thus pose the problem of classifying all possible "function groups" that can be defined as submersion maps $T^d(M) \to M''$ as explained in Section 6. This is clearly an interesting way to generate "function groups".

Here is one important example. Suppose that G is a Lie group. Let $\underset{\sim}{G}$ be its Lie algebra, identified with the vector space of left-invarient vector fields on G. The tangent bundle to G is then isomorphic to

$$G \times \underset{\sim}{G} \quad .$$

$T^d(M)$ is isomorphic to

$$G \times \underset{\sim}{G}^d \quad .$$

It is readily seen that the Cartesian projection

$$\phi\colon G \times \underset{\sim}{G}^d \to \underset{\sim}{G}^d$$

has the following property:

> ϕ is a homomorphism between the cosymplectic structure on $\underset{\sim}{G}^d$ defined by the Lie algebra structure and the cosymplectic structure defined by the natural symplectic structure on $T^d(G) \approx G \times \underset{\sim}{G}^d$.

8. LIE'S CANONICAL FORM FOR A COSYMPLECTIC STRUCTURE OF CONSTANT RANK. THE CO-CHARACTERISTIC FOLIATION

I refer now to the material in Chapter 5, particularly Theorem 23, in Volume 2 of Lie's "Transformationsgruppen". I will present a modernized version of some of this material, together with a new concept, the _co-characteristic foliation_.

Let M be a manifold, $\omega\colon F^1(M) \times F^1(M) \to F(M)$ a tensorial form which defines a cosymplectic structure. For each $p \in M$, the value of ω at p is a form $M_p^d \times M_p^d \to R$ The _rank_ of ω at p is the rank of this form. In local coordinates (x^i), $1 \leq i,j \leq m$, for M, with

$$\omega(dx^i, dx^j) = \omega^{ij} \quad ,$$

the rank at p is then the rank of the skew-symmetric matrix

$$(\omega^{ij}(p)) \quad .$$

Suppose that this integer is constant as p ranges over M.

$$\omega \text{ defines } F(M) \text{ as a Lie algebra.}$$

Let $\mathscr{C}$ be the center of this Lie algebra, i.e., the set of $f \varepsilon F(M)$ such that

$$\{f, F(M)\} \equiv \omega(df, dF(M)) = 0$$

(Lie calls these "ausgezeichnete Functionen".)

For each $f \varepsilon F(M)$, let A_f be the vector field on M such that

$$A_f(f_1) = \{f, f_1\}$$
$$\text{for each } f, f \ \varepsilon \ F(M) \quad .$$

Then, for $f_1, f_2, f_3 \ \varepsilon \ F(M)$,

$$\left[A_{f_1}, A_{f_2}\right](f_3) \qquad A_{f_1}\left(A_{f_2}\right)(f_3) - A_{f_2}A_{f_1}(f_3)$$

$$= A_{f_1}(\{f_2, f_3\}) - A_{f_2}(\{f_1, f_3\})$$

$$= \{f_1, \{f_2, f_3\}\} - \{f_2, \{f_1, f_3\}\}$$

= , using the Jacobi identity,

$$\{\{f_1, f_2\}, f_3\}$$

$$= A_{\{f_1, f_2\}}(f_3) \quad .$$

Hence,

$$\left[A_{f_1}, A_{f_2}\right] = A_{\{f_1, f_2\}} \tag{8.1}$$

This identity tells us that the map

$$f \rightarrow A_f \tag{8.2}$$

of $F(M) \rightarrow$ (vector fields on M) is a Lie algebra homomorphism. Clearly,

$$\mathscr{C} \text{ is the kernel of this homomorphism} \tag{8.3}$$

Let $\mathscr{L}$ be the Lie algebra of vector fields spanned by the A_f, $f \varepsilon F(M)$. For $p \varepsilon M$,

$$\mathscr{L}(p) = \{A_f(p)\colon f \varepsilon F(M)\} \subset M_p$$

Then,

$$\dim \mathscr{L}(p) = \operatorname{rank} \omega \quad . \tag{8.4}$$

Hence,

$$[\mathscr{L}, \mathscr{L}] \subset \mathscr{L} \quad ,$$

it defines a <u>foliation</u> of M by submanifolds, called the <u>co-characteristic foliation</u>.

Let $\theta \varepsilon M_p^d$ be orthogonal to $\mathscr{L}(p)$. We see that:

$$\theta \text{ is a characteristic for } \omega \quad ,$$

$$\omega(\theta, M_p^d) = 0 \tag{8.5}$$

Let $L(p)$ be the leaf of the foliation $\mathscr{L}$ passing through p.

We see that:

$$L(p)^d = M_p^d/(\text{characteristics of } \omega) \quad .$$

Hence, ω passes to the quotient to define a cosymplectic structure on $L(p)$, denoted by $\overline{\omega}$. This form has no nonzero characteristics, i.e., $\overline{\omega}$ <u>defines a symplectic structure on each submanifold</u> $L(p)$. <u>Also</u>, <u>the functions on</u> M <u>which are constant along the leaves of</u> $\mathscr{L}$ <u>are precisely the elements of</u> $\mathscr{C}$.

What we now call Darboux' theorem for symplectic manifolds (but which probably is also due to Lie?) now implies that there is a coordinate system for $L(p)$,

$$(x^i, y_j) \quad , \qquad 1 \le i,j \le r$$

$$2r = \dim L(p) \quad ,$$

such that

$$\begin{aligned} \{x^i, y_j\} &= \delta^i_j \\ \{x^i, x^j\} &= 0 = \{y_i, y_j\} \quad . \end{aligned} \tag{8.6}$$

As p varies, we obtain a set of functionally independent functions defined locally on M which also satisfy (8.6). Putting these together with a maximal functionally independent set (z^a), $1 \le a \le s$, of functions in $\mathscr{C}$ provides a local coordinate system

$$(x^i, y_j, z^a)$$

for M such that, in addition to (8.5), the following conditions are satisfied:

$$\{z^a, x^i\} = \{z^a, y_i\} = 0 \tag{8.7}$$

(8.6) and (8.7) exhibit what Lie calls the canonical form for the cosymplectic structure.

Notice that we obtain a good deal of geometric information from this argument. M is foliated by submanifolds on which is induced a symplectic structure.

9. THE CO-CHARACTERISTIC FOLIATION ON THE DUAL SPACE TO A LIE ALGEBRA

Let G be a Lie group, $\underset{\sim}{G}$ its Lie algebra, $\underset{\sim}{G}^d$ its dual space. We know that $\underset{\sim}{G}^d$ has a cosymplectic structure. We can readily calculate its co-characteristic foliation.

Let $(\underset{\sim}{g}^i)$, $1 \leq i,j \leq n$, be a basis for $\underset{\sim}{G}$ with

$$[\underset{\sim}{g}^i, \underset{\sim}{g}^j] = c_k^{ij} \underset{\sim}{g}^k \tag{9.1}$$

Let $y^i \in F(\underset{\sim}{G}^d)$ be the function on $\underset{\sim}{G}^d$ such that

$$y^i(\theta) = \theta(\underset{\sim}{g}^i) \tag{9.2}$$

for $\theta \in \underset{\sim}{G}^d$.

Then, the cosymplectic structure satisfies:

$$\{y^i, y^j\} = c_k^{ij} y^k \quad . \tag{9.3}$$

Hence,

$$\omega(dy^i, dy^j) = c_k^{ij} y^k \quad . \tag{9.4}$$

Suppose that:

$$\theta = a_i dy^i$$

is a co-characteristic vector of ω, i.e.,

$$0 = \omega(\theta, dy^j)$$

$$= a_i c_k^{ij} y^k \tag{9.5}$$

Let $(\underset{\sim}{g}_i)$ be the basis of $\underset{\sim}{G}^d$ which is dual to the basis $(\underset{\sim}{g}^i)$ of $\underset{\sim}{G}$, i.e.,

$$\underset{\sim}{g}_i(\underset{\sim}{g}^j) = \delta_i^j \quad . \tag{9.6}$$

Set:

$$p = y^k \underset{\sim}{g}_k \tag{9.7}$$

$$\underset{\sim}{g} = a_i \underset{\sim}{g}^i \quad . \tag{9.8}$$

Condition (9.5) means that:

$$0 = a_i \underset{\sim}{g}_k([\underset{\sim}{g}^i, \underset{\sim}{g}^j]) y^k$$

$$= p([\underset{\sim}{g}, \underset{\sim}{g}^j])$$

$$= \mathrm{adj}^d(\underset{\sim}{g})(p)(\underset{\sim}{g}^j) \tag{9.9}$$

where

$$\mathrm{adj}^d : \underset{\sim}{G} \rightarrow L(\underset{\sim}{G}^d)$$

is the dual of the adjoint representation of $\underset{\sim}{G}$. Hence, we have proved the following:

Theorem 9.1. Let p be a point of $\underset{\sim}{G}^d$. Identify the cotangent space to $\underset{\sim}{G}^d$ at p with $\underset{\sim}{G}$. Then, a vector $\underset{\sim}{g} \in \underset{\sim}{G}$ determines a characteristic co-vector for the cosymplectic form ω if and only if

$$\mathrm{adj}^d(\underset{\sim}{g})(p) = 0 \tag{9.10}$$

Further, the Casimir operators (i.e., the center) of $U(\underset{\sim}{G})$ are in $\mathscr{C}$, the center of the Poisson bracket algebra on $F(\underset{\sim}{G}^d)$.

10. COSYMPLECTIC STRUCTURES ON QUOTIENTS BY GROUPS OF AUTOMORPHISMS

Let M be a manifold with a cosymplectic structure ω. If $\phi: M \to M'$ is a map, there is no general reason that there should be a cosymplectic structure ω' on M' such that ϕ is a homomorphism between ω and ω'. However, in one important case there will be, as covered by the following result.

Theorem 10.1. Suppose that:

a) $\phi: M \to M'$ is a submersion map

b) There is a group G acting on M consisting of automorphisms of ω which act transitively on the fibers of ϕ.

Then, there is a cosymplectic structure ω' on M' such that ϕ is a homomorphism between M and M'.

Proof. For $p \in M$, let $\omega(p)$ denote the value of ω at p. As a skew-symmetric bilinear form on M_p^d, it is an element of

$$M_p \wedge M_p$$

the skew-symmetric tensor product of the tangent space with itself. Let

$$\phi_*: M_p \to M'_{\phi(p)}$$

be the linear map ϕ induces on tangent vectors. Its skew-symmetric tensor product is defined:

$$\phi_* \wedge \phi_*: M_p \wedge M_p \to M'_{\phi(p)} \wedge M'_{\phi(p)} \quad .$$

It is readily seen (exercise!) that the condition that ϕ define a homomorphism between ω and ω' is that:

$$(\phi_* \wedge \phi_*)(\omega(p)) = \omega'(\phi(p)) \tag{10.1}$$

for <u>all</u> $p \in M$.

To define a $\omega': p' \to \omega'(p') \in M'_{p'} \wedge M'_{p'}$ satisfying (10.1). proceed as follows. For each point $p \in \phi^{-1}(p')$, set:

$$\omega'(p') = (\phi_* \wedge \phi_*)(\omega(p)) \tag{10.2}$$

If $g \in G$, then g maps p onto another point of the fiber. Since g preserves ω, we have:

$$g_*(\omega(p)) = \omega(gp) \quad .$$

In particular,

$$\phi_* \wedge \phi_*(\omega(p)) = \phi_* \wedge \phi_*(\omega(gp)) \quad .$$

This shows that $\omega'(p')$ as <u>defined</u> by (10.2) is independent of the $p \in \phi^{-1}(p')$ chosen. Proof that $p' \to \omega'(p')$ depends smoothly on p', i.e., that this defines a cosymplectic structure, is left to the reader as an exercise.

Here is another useful way of looking at this construction.

$$\phi^*: F(M') \to F(M)$$

is one-one (since ϕ is a submersion), hence defines $\phi^*(F(M'))$ as a subalgebra of $F(M)$. Let $\{ \ , \ \}$ denote the Poisson bracket operation on $F(M)$ defined by the cosymplectic structure ω. The assumption that G acts on M so as to map fibers of ϕ onto fibers, <u>and</u> that G acts transitively on the fibers means that:

> $\phi^*(F(M'))$ are the set of functions on M which are invariant under the action of G.

Call these invariants

$$\mathscr{I}(F(M),G) \quad .$$

Since G acts on a group of automorphisms of the cosymplectic structure, we have:

$$\{\mathscr{I}(F(M),G), \mathscr{I}(F(M),G)\} \subset \mathscr{I}(F(M),G) \quad ,$$

i.e., the Poisson bracket on M induces a Poisson bracket on $\mathscr{I}(F(M),G)$, hence also on $\phi^*(F(M'))$, hence also on F(M'), since ϕ^* is an isomorphism. Now, this Poisson bracket operation on F(M') defines a cosymplectic structure ω' on M', as its "symbol". (See [11].) This ω' is obviously the structure provided by Theorem 4.1.

BIBLIOGRAPHY

1. V. Arnold, Ann. Inst. Grenoble, 16 (1966), 319-361.

2. P. Lax, "Integrals of nonlinear equations of evolution and solitary waves", Comm. Pure App. Math. 21 (1968), pp. 467-490.

3. R. Hermann, Gauge Fields and Cartan-Ehresmann Connections (Interdisciplinary Mathematics, Vol. 10), Math Sci Press, Brookline, Mass., 1975.

4. R. Hermann, Topics in the Mathematics of General Relativity (Interdisciplinary Mathematics, Vol. 5), Math Sci Press, Brookline, Mass., 1973.

5. B. Kostant, "Quantization and Unitary Representations", Lecture Notes in Math., Vol. 170, Springer-Verlag, 87-208.

6. E. Whittaker, Analytical Dynamics, Cambridge University Press, London, 1969.

7. M. Toda, "Wave propagation on anharmonic lattices", J. Phys. Soc. Japan 23 (1967), 501-506.

8. H. Flaschka, "The Toda Lattice", I&II, Phys. Rev. B9 (1974), 1924-1925; Prog. Theor. Phys. 51 (1974), 703-716.

9. J. Moser, "Finitely many mass points on the line under the influence of an exponential potential--an integrable system", in Lecture Notes in Physics, Vol. 38, Springer-Verlag, 1975.

10. Sophus Lie, Theorie der Transformationsgruppen, Chelsea Publ. Co., New York, 1970.

11. R. Hermann, Geometry, Physics and Systems, Marcel Dekker, New York, 1973.

THE DIFFERENTIAL-GEOMETRIC FOUNDATION OF THE INVERSE SCATTERING TECHNIQUE, III

THE ALGEBRAIC BASIS FOR THE GELFAND-LEVITAN TECHNIQUE; GRADED ASSOCIATIVE ALGEBRAS; AND COSYMPLECTIC STRUCTURES

ABSTRACT

The Gelfand-Levitan equations of Inverse Scattering Theory are interpreted in a Lie group theoretic way. A general algebraic framework--involving graded associative algebras--is presented as the appropriate algebrazation of both the Gelfand-Levitan technique and the Toda lattice situation. This framework is capable of extension in many directions, especially towards infinite dimensional situations.

1. INTRODUCTION

As we have seen, in group theoretic terms, the basic general equation of the theory of solitons is:

$$\frac{da}{dt} = [a,\alpha(a)] \quad , \tag{1.1}$$

where $\mathcal{G}$ is a real Lie algebra, $\mathcal{A}$ is a subset of $\mathcal{G}$, $\alpha\colon \mathcal{A} \to \mathcal{G}$ is a map, and (1.1) is to be solved for a curve $t \to a(t)$ in $\mathcal{A}$.

Now, (1.1) can be "solved" as follows. Let G be a group whose Lie algebra is $\mathcal{G}$. Let $\mathrm{ad}(g)\colon \mathcal{G} \to \mathcal{G}$ denote the adjoint linear representation of G in $\mathcal{G}$. ($\mathcal{G}$ is identified with the one-parameter subgroups of G. $\mathrm{Ad}\ g$ acts via conjugation on such one-parameter subgroups.) Then, Equation (1.1) can be "solved" in the following form:

$$a(t) = \mathrm{Ad}(g(t))(a(0)) \quad , \tag{1.2}$$

where $t \to g(t)$ is a curve in G.

This transforms the problem of solving (1.1) into the problem of finding the curve $t \to g(t)$ in G. Now, for an arbitrary curve $t \to g(t)$ on G, the constraints that $a(t) \in \mathcal{A}$ will not be satisfied. The main problem--which is still unsolved in reasonable generality--is to see what these constraints mean. In the main cases where we do understand what is going on (Toda lattice, Korteweg-de Vries, ...), these constraints are closely linked to the specific mechanism for solving the traditional Inverse Scattering Problem developed by Gelfand and Levitan [1], Marchenko [2], Kay and Moser [3,4,5]. In fact, I shall recast their ideas into a more Lie-algebraic form, covering also the more recent "discrete" versions of the Inverse Scattering Technique developed by Case and Kac [6]. The general idea seems to be to express the constraints by requiring that the curve $t \to g(t)$ lie in a subgroup of G. In the situations considered in the literature, this subgroup seems to be solvable. This leads to various interesting ramifications which I shall develop in this and succeeding papers. In particular, it seems to me that this solvability property is the key one for understanding the meaning of the Gelfand-Levitan technique (and for possible generalizations!) It also is directly related to the "solubility" (in the classical sense) of the equations (1.1). In fact, it is this solvable group which provides the "angle" variables for (1.1) interpreted as a Hamiltonian system. (The eigenvalues when a is described by matrices, provide the "action" variables.)

2. THE GELFAND-LEVITAN IDEA

Let us begin with a finite dimensional situation. Let V be a finite dimensional real vector space. Let $L(V)$ denote the space of linear maps: $V \to V$. Let

$$A_0, A_1 \in L(V)$$

be two elements, which are both diagonalizable. Suppose they have the same eigenvalues, $\lambda^1,\dots,\lambda^n$. Let

$$(v_i),(u_i) \quad , \qquad 1 \leq i,j = n = \dim V$$

be their eigenvectors, with corresponding eigenvalues.

Let us propose to find a linear map

$$B \in L(V)$$

such that:

$$B(v_i) = u_i \tag{2.1}$$

$$\text{for } i = 1,\dots,n \quad .$$

In other words, V maps the eigenvectors for A_0 onto those for A_1, preserving the corresponding eigenvalues. Thus,

$$A_1B(v_i) = A_1(u_i) = \lambda^i u_k \quad \text{(no summation)}$$

$$= \lambda^i(v_i) = \alpha(\lambda^i v_i) = A_0(v_i) \quad .$$

Since the v_i form a basis for V, we must have:

$$A_1B = BA_0 \quad , \tag{2.2}$$

i.e., α is an <u>intertwining operator</u> between A_0 and A_1. We shall pose the <u>Gelfand-Levitan problem</u> as finding maps B satisfying (2.2).

Suppose in addition, that (x_i) is a fixed basis for V. Suppose that:

$$v_i = a_i^j x_j \tag{2.3}$$

$$u_i = c_i^j v_j \quad , \tag{2.4}$$

with matrices (a_i^j), (c_i^j). (The summation convention is now in force.) Thus, if

$$u_i = v_j \tag{2.5}$$

then:

$$c_i^j = B_i^k a_k^j \quad . \tag{2.6}$$

The Gelfand-Levitan idea is to look for B so that:

$$u_i = \sum_{j=1}^{i} B_i^j v_j \quad , \tag{2.7}$$

i.e.,

$$u_1 = B_1^1 v \quad ,$$

$$u_2 = B_2^2 v_2 + B_2^1 v_1$$

$$u_3 = B_3^3 v_3 + B_3^2 v_2 + B_3^1 v \quad ,$$

etc.

This requires that

$$B_i^j = 0 \ , \quad \text{for } j > i \quad .$$

The (B_i^j) are a discrete analog of what analysts call a <u>Volterra kernel</u>. B is a <u>Volterra operator</u>. Group theoretically, it lies in the maximal solvable subgroup of $GL(n,R)$.

Lie group theoretically, we see that this means that:

> A_1 lies on the orbit $\text{Ad } S(A_0)$ of a maximal solvable subgroup of $GL(V)$ acting on A_0.

There is no guarantee that such an intertwining B exists. When they do, interesting things happen. We shall see in the next section what algebraic structure is involved.

These ideas can be carried over to infinite dimensions in many ways. However, one must take care with the meaning of "solvability". One way--which was that followed by Gelfand and Levitan, and almost all authors following them--is to substitute the notion of "integral" operator with triangular kernel", or <u>Volterra operator</u>. For example, let us refer to the original example [1].

Set:

$V = C^{\infty}$, real-valued function on the real number interval $[0,\infty)$, with certain boundary conditions which will be specified later.

Denote the coordinate on $[0,\infty)$ by x. Set:

$$A_1 = \frac{d^2}{dx^2} - q(x)$$

$$A_0 = \frac{d^2}{dx^2} \quad .$$

Look for $\alpha \colon V \to V$ so that

$$\alpha(v)(x) = v(x) + \int_0^x K(x,y)v(y)\, dy \quad , \tag{2.8}$$

i.e.,

$$\alpha = 1 + N \tag{2.9}$$

with

$$N(v)(x) = \int_0^x K(x,y)v(y)\,dy \quad . \tag{2.10}$$

The <u>Gelfand-Levitan integral equation</u> is:

$$f(x,y) + \int_0^x f(x,y)K(x,y)\,dy + K(x,y) = 0 \tag{2.11}$$

It can be written as an operator equation of the following form:

$$F + FN^T + N = 0 \quad , \tag{2.12}$$

where F is a linear map determined by the spectral data of A_1, and N^T denotes the transpose operator to N with respect to the usual inner product

$$\langle v_1, v_2 \rangle = \int_0^\infty v_1(x)v_2(x)\,dx \tag{2.13}$$

on V.

We shall now leave the complicated infinite dimensional situation in order to analyze things more completely in the finite dimensional context.

3. JACOBI ELEMENTS OF GRADED ASSOCIATIVE ALGEBRAS

We work within an algebraic setting which will ultimately prove to be relevant to our goal of analyzing the Gelfand-Levitan ideas. Let $\mathcal{A}$ be an associative algebra over the real numbers with a unit element 1. $\mathscr{A}$ is graded if it is (as a vector space) the direct sum

$$\cdots + \mathscr{A}^{-2} + \mathscr{A}^{-1} + \mathscr{A}^{0} + \mathscr{A}^{1} + \mathscr{A}^{2} \cdots$$

of a sequence of linear subspaces such that:

$$\mathscr{A}^{j}\mathscr{A}^{k} \subset \mathscr{A}^{j+k} \tag{3.1}$$

$$\text{for} \quad -\infty < j,k \leq \infty \quad .$$

We can also consider $\mathscr{A}$ to be a Lie algebra via commutation

$$[a,b] = ab - ba \quad .$$

Let

$$\mathscr{S} = \mathscr{A}^{0} + \mathscr{A}^{1} + \mathscr{A}^{2} + \cdots$$

$$S = \text{elements of } \mathscr{S} \text{ which are invertible.}$$

Definition. An element $a \in \mathscr{A}$ is a Jacobi element if it is of the form

$$a = a_{-1} + a_{0} + a_{1} \quad .$$

Our fundamental problem is to determine when two Jacobi elements a and $b = b_{-1} + b_{0} + b_{1}$ are conjugate under the action of S, i.e., $b = cac^{-1}$, for $c \in S$. Thus, if

$$c = c_{0} + c_{1} + c_{2} + \cdots \quad ,$$

we must have:

$$bc = ca \quad ,$$

or

$$(b_{-1}+b_0+b_1)(c_0+c_1+c_2+\cdots) = (c_0+c_1+\cdots)(a_{-1}+a_0+a_1)$$

Equate terms of equal gradation:

$$b_{-1}c_0 = c_0a_{-1} \tag{3.2}$$

$$b_0c_0 + b_{-1}c_1 = c_0a_0 + c_1a_{-1} \tag{3.3}$$

$$b_1c_0 + b_0c_1 + b_{-1}c_2 = c_0a_1 + c_1a_0 + c_2a_{-1} \tag{3.4}$$

and so forth.

(3.2) and (3.3) can be combined as follows

$$b_0c_0 + (c_0a_{-1}c_0^{-1})c_1 = c_0a_0 + c_1a_{-1} \quad . \tag{3.5}$$

Thus, the first "obstruction" is the solvability of (3.5) for c_1. The simplest case is that where the equation

$$(c_0a_{-1}c_0^{-1})x - xa_{-1} = y \tag{3.6}$$

can be solved for $x \in \mathscr{A}^1$ whatever the $y \in \mathscr{A}^0$. We can continue in this way for $y \in \mathscr{A}^1, \mathscr{A}^2, \ldots$.

<u>Theorem 3.1</u>. Suppose that a,b are two Jacobi elements of such that there exists $c_0 \in \mathscr{A}^0$, such that c_0^{-1} exists, and (3.2) is satisfied. Suppose that given $y \in \mathscr{A}^1, \mathscr{A}^1, \ldots$ there is $x \in \mathscr{A}^1, \mathscr{A}^2, \ldots$ satisfying (3.6). Then, there is an $s \in S$ such that

$$b = sas^{-1} \quad .$$

In fact, it is easy to see that condition (3.6) can never be satisfied in finite dimensional situations. (This is one good reason for the Gelfand-Levitan phenomenon being primarily associated with differential or difference equation situations!) Instead, we shall look for further conditions, motivated by the Lax equations.

4. THE LAX EQUATIONS FOR THE JACOBI ELEMENTS

Continue with the notation of Section 3. However, suppose that $\mathscr{A}$ is a finite dimensional vector space. Let

$$\mathscr{A} = \mathscr{A}_{-1} + \mathscr{A}_0 + \mathscr{A}_1 \tag{4.1}$$

be a Jacobi element. Set:

$$\alpha(a) = \mathscr{A}_{-1} - \mathscr{A}_1 . \tag{4.2}$$

Then,

$$\begin{aligned}[\alpha(a),a] &= [a_{-1} - a_1, a_{-1} + a_0 + a_1] \\ &= [a_{-1},a_0] - [a_1,a_0] + 2[a_{-1},a_1] \end{aligned} \tag{4.3}$$

Now let $\underset{\sim}{S}$ denote the set of non-negative degree elements of $\mathscr{A}$, i.e.,

$$\underset{\sim}{S} = \mathscr{A}^0 + \mathscr{A}^1 + \mathscr{A}^2 + \cdots$$

Theorem 4.1. For each Jacobi element a of $\mathscr{A}$, there is a $s \in \underset{\sim}{S}$ such that:

$$[\alpha(a), a] = [s, a] \tag{4.4}$$

In fact, s can be taken in the following form:

$$s = -a_0 - 2a_1 \quad . \tag{4.5}$$

Proof. With s given by (4.5), we have:

$$\begin{aligned}[s,a] &= [a, a_0+2a_1] \\ &= [a_{-1}+a_0+a_1, a_0+2a_1] \\ &= [a_{-1},a_0] + 2[a_{-1},a_1] + 2[a_0,a_1] + [a_1,a_0] \\ &= [a_{-1},a_0] + 2[a_{-1},a_1] - [a_1,a_0] \quad ,\end{aligned}$$

which is precisely the right hand side of (4.3)

We can now give the Lie group and geometric interpretation of this result.

Theorem 4.2. Let S be the Lie group generated by $\underset{\sim}{S}$. Let $\mathscr{J}$ be the vector space of Jacobi elements of $\mathscr{A}$. Let X be the vector field on $\mathscr{J}$ defined by the following formula:

$$X(z) = [\alpha(a), a] \tag{4.6}$$

for $a \in \mathscr{J}$.

Then, X is always tangent to the orbits of S acting on $\mathscr{J}$.

<u>Corollary to Theorem 4.2</u>. Let $t \to a(t)$ be a curve on $\mathscr{J}$ that is, a solution of the equations

$$\frac{da}{dt} = [\alpha(a), a] \quad .$$

Then, there is a curve $t \to s(t)$ on S such that

$$a(t) = s(t)(a(0)) \quad . \tag{4.3}$$

Equations (4.3) are basically <u>Gelfand-Levitan</u>, specialized to this situation!

<u>Remarks</u>. a) To specialize to the Toda lattice, proceed as follows

$\mathscr{A}$ = $n \times n$ matrices

$\mathscr{A}^0$ = diagonal matrices

$\mathscr{A}^1$ = matrices with zeros everywhere except for the first super-diagonal

$\mathscr{A}^{-1}$ = transposes of $\mathscr{A}^1$

$\mathscr{A}^2 = \mathscr{A}^1\mathscr{A}^1$; $\mathscr{A}^{-2} = \mathscr{A}^{-1}\mathscr{A}^{-1}$,

etc.

For example, for $n = 3$,

$$\mathscr{A}^0 = \left\{ \begin{pmatrix} x_1 & 0 & 0 \\ 0 & x_2 & 0 \\ 0 & 0 & x_3 \end{pmatrix} \right\}$$

$$\mathscr{A}^1 = \left\{\begin{pmatrix} 0 & y_1 & 0 \\ 0 & 0 & y_2 \\ 0 & 0 & 0 \end{pmatrix}\right\}$$

$$\mathscr{A}^{-1} = \left\{\begin{pmatrix} 0 & 0 & 0 \\ z_1 & 0 & 0 \\ 0 & z_2 & 0 \end{pmatrix}\right\}$$

$$\mathscr{J} = \left\{\begin{pmatrix} x_1 & y_1 & 0 \\ z_1 & x_2 & y_2 \\ 0 & z_2 & x_3 \end{pmatrix}\right\}$$

We see that Theorems 4.1 and 4.2 sum up the main properties of the Toda lattice equations, as interpreted in the work of Flaschka [7].

b) Although strictly speaking, these are not true in infinite dimensions, since the Lie theory we have used does not generalize, the "formal" part goes over. Thus, (4.3) will hold, with $s(t)$ given by a formal power series:

$$s(t) = s_0 + s_1 t + \cdots \tag{4.4}$$

The purpose of a "generalized Gelfand-Levitan theory" would be to provide some proper functional analysis setting for the infinite dimensional situation; perhaps to interpret the series (4.4) in a suitable way. (Much as functional analysis provides us with the conceptual tools to understand the exponential series $e^{At} = 1 + At + \frac{1}{2}A^2t^2 + \cdots$ in infinite dimensional situations. In fact, this whole business can be looked on as a generalization of this situation (reducing to it if β = constant map), hence, I believe that most of the "semigroup-theoretical" functional analysis tools will carry over in some form or other.)

c) Much more general graded associative algebras $\mathscr{A}$ (even finite dimensional ones) can be envisaged for those leading to the Toda lattice. In fact, suppose V is a finite dimensional vector space which is a direct sum of linear subspaces

$$V = V_{-m} + V_{-m+1} + \cdots + V_n$$

Set:

$$\mathscr{A} = L(V) \equiv \text{space of linear maps } V \to V$$

$$\mathscr{A}^j = \{a \in \mathscr{A} : a(V^k) \subset V^{k+j} \text{ for all } k\}$$

The assignment $j \to \mathscr{A}^j$ then provides us with a gradation.

d) The algebraic set-up described above is so simple that it can obviously be widely generalized. For example, it would be enough to postulate a Lie algebra structure for $\mathscr{A}$, together with a gradation by linear subspaces $\mathscr{A}^j$ such that

$$[\mathscr{A}^j, \mathscr{A}^k] \subset \mathscr{A}^{j+k}$$

(Note that this should <u>not</u> be called a "graded Lie algebra"--these are different objects!) Obviously, what is called for is a deeper, more systematic analysis of ordinary differential equations whose right hand sides are quadratic. What should be done is to provide an algebraic classification of when these equations exhibit the "soliton" and "inverse scattering" phenomenon.

5. THE LAX EQUATION AS A VECTOR FIELD ON THE LIE ALGEBRA

As one might expect, the ramifications of these ideas are widespread. (This seems to be one of the most striking and, no doubt important, features of soliton theory.) Here is another, more "intrinsic" (from the point of view of Lie group theory) way of thinking of this material.

Let G be a Lie group with $\mathcal{G}$ its Lie algebra. Let $\mathcal{A}$ be a subset of $\mathcal{G}$, and let

$$\alpha\colon \mathcal{A} \to \mathcal{G}$$

be a map. (Do not necessarily assume that $\mathcal{A}$ is a linear subspace, nor that α is linear.) Let G act on $\mathcal{G}$ via the adjoint representation. A solution $t \to a(t)$ of the Euler-Arnold-Lax equation

$$\frac{da}{dt} = [\alpha(a),a] \tag{5.1}$$

then lies on the orbit of G acting on the point $a(0)$ of $\mathcal{A}$.

The properties of this orbit $\mathrm{Ad}\, G(a(0))$ will, of course, depend on the properties of $a(0) \in \mathcal{G}$, and of the isotropy subgroup of G acting on a(0), which is:

$$G^{a(0)} = \{g \in G: \text{Ad } g(a(0)) = a(0)\} \quad .$$

One may expect that at least some of the properties of the differential equations (5.1) (and of the physical systems which they model) depend in a qualitative way on the nature of $G^{a(0)}$ and the orbit, which is the coset space $G/G^{a(0)}$. We are interested in the following equation.

> When do the solutions of (5.1) lie on the orbits of subgroups S of G?

In order to approach these questions, let us make the following assumption:

> The linear map $\text{Ad } a(0): \mathcal{G} \to \mathcal{G}$ is diagonalizable and has only <u>real</u> eigenvalues. (5.2)

With this assumption, let:

$\mathcal{G}^+$ = linear subspace of $\mathcal{G}$ spanned by the eigenvectors of Ad a(0) with positive eigenvalues

$\mathcal{G}^-$ = linear subspace of $\mathcal{G}$ spanned by the eigenvalues of Ad a(0) with negative eigenvalues

$\mathcal{G}^0$ = linear subspace of $\mathcal{G}$ spanned by eigenvectors of zero eigenvalue, i.e., $\mathcal{G}^0$ = Lie algebra of $G^{a(0)}$.

Notice that

$$[\mathscr{G}^{+}, \mathscr{G}^{+}] \subset \mathscr{G}^{+} \tag{5.3}$$

$$[\mathscr{G}^{-}, \mathscr{G}^{-}] \subset \mathscr{G}^{-} \tag{5.4}$$

$$[\mathscr{G}^{0}, \mathscr{G}^{\pm}] \subset \mathscr{G}^{\pm} \tag{5.5}$$

$$\underset{\sim}{G} = \text{direct sum as vector space of } \underset{\sim}{G}^{+} + \underset{\sim}{G}^{-} + \underset{\sim}{G}^{0}$$

Let G^{+}, G^{-} be the connected subgeoups of G generated by the Lie subalgebra $\mathscr{G}^{+}$, $\mathscr{G}^{-}$. Consider the coset space

$$G/G^{a(0)} = \text{Ad } G(a(0))$$

Now, map

$$\phi\colon G^{-1} \times G^{+} \to G/G^{a(0)}$$

as follows

$$\phi(g^{-}, g^{+}) = \text{Ad } g^{-} \text{ Ad } g^{+}(a(0)) \tag{5.6}$$

It follows from (5.3)-(5.5) that this map sends the tangent space at (1,1) (where "1" denotes the identity-element of groups) isomorphically into the tangent space to $G/G^{a(0)}$. Hence, the image of ϕ at least contains an open subset of $G/G^{a(0)}$. (In fact, in typical situations ϕ is a diffeomorphism, or at least the image $\phi(G^{-} \times G^{+})$ is a "large" subset in the sense that its complement is obtained in a "Zariski closed" subset of $G/G^{a(0)}$.)

Further, let us make the following assumption about $\underset{\sim}{G}$:

> There is a non-degenerate, symmetric bilinear form on $\underset{\sim}{\mathcal{G}}$ which is invariant under the adjoint representation.

<u>Remark</u>. This condition is equivalent to $\underset{\sim}{G}$ being a reductive Lie algebra, i.e., the direct sum of a semisimple Lie algebra and a abelian Lie algebra.

As explained in Interdisciplinary Mathematics, Volume 6, Kostant's construction [8] now provides a symplectic structure for the orbit

$$\mathrm{Ad}\, G(a(0)) \quad .$$

The orbits $G^+(a(0))$, $G^-(a(0))$ are Lagrangian submanifolds with respect to this symplectic structure. There is then a fibration

$$G/G^{a(0)} \to G/G^+ \quad . \tag{5.7}$$

This is a <u>Lagrangian fibration</u>, i.e., its fibers are Lagrangian submanifolds with respect to the symplectic form on $G/G^{a(0)}$.

Let $X_\alpha: \mathcal{G} \to \mathcal{G}$ be the vector field on $\mathcal{G}$ defined by the following formula

$$X_\alpha(Y) = [\alpha(Y,Y] \quad .$$

The orbit curves of X_α are then the solutions of the Lax equation:

$$\frac{dY}{dt} = [\alpha(Y), Y] \quad . \tag{5.8}$$

Theorem 5.1. If the linear map α is symmetric with respect to the bilinear form on $\mathcal{G}$, then the vector field X_α, restricted to each orbit of Ad G, is an infinitesimal symplectic automorphism of Kostant's symplectic structure on the orbit.

We shall actually prove this result by working with the cosymplectic structure on $\mathcal{G}^d$, to which we now turn.

6. THE EULER-LAX VECTOR FIELD ON THE DUAL TO THE LIE ALGEBRA AND ITS ACTION AS AN INFINITESIMAL SYMPLECTIC AUTOMORPHISM ON KOSTANT'S SYMPLECTIC STRUCTURE

Let $\mathcal{G}$ denote a real Lie algebra, with $\mathcal{G}^d$ its dual. Let $n = \dim \mathcal{G}$. Choose indices as $1 \leq i,j \leq n$ and the summation convention. Let $F(\mathcal{G}^d)$ denote the C^∞, real-valued functions on $\mathcal{G}^d$. Let $y^i \in F(\mathcal{G}^d)$ be a basis for linear functions on $\mathcal{G}^d$. Let

$$\{\ ,\ \}: F(\mathcal{G}^d) \times F(\mathcal{G}^d) \to F(\mathcal{G}^d)$$

denote the Poisson bracket operation defined by the cosymplectic structure on $\mathcal{G}^d$. Then,

$$\{y^i, y^j\} = c_k^{ij} y^k \quad , \tag{6.1}$$

where c_k^{ij} are the structure constants of $\mathcal{G}$. (Thus, since $(\mathcal{G}^d)^d = \mathcal{G}$, the y^j can be identified with a basis for $\mathcal{G}$, say, X^i, and $[X^i,X^j] = c_k^{ij}X^k$.) The cosymplectic form ω is defined as follows:

$$\omega(dy^i,dy^j) = c_k^{ij}y^k \quad . \tag{6.2}$$

Now, set

$$h = \frac{1}{2} b_{ij}y^iy^j \quad , \tag{6.3}$$

where (b_{ij}) are the components of a symmetric bilinear form on $\mathscr{G}^d$. We compute the vector field A_h on $\mathscr{G}^d$ defined by h

$$A_h(f) = \{h,f\} = \omega(dh,df) \tag{6.4}$$

$$\text{for } f \in F(\mathscr{G}^d)$$

A_h is automatically an infinitesimal automorphism of the cosymplectic structure on $F(\mathscr{G}^d)$. (This is just essentially what the Jacobi identity for the Poisson bracket says.) In particular, if A_h is tangent to the orbits of Ad^dG (the dual of the adjoint representation of G) acting on $\mathscr{G}^d$, then <u>automatically</u> A_h preserves Kostant's symplectic structure on these orbits.

Let us therefore see whether A_h is tangent to these orbits.

$$dh = b_{ij}y^idy^j$$

$$A_h(y^k) = \omega(dh,dy^k) = \omega(b_{ij}y^idy^j,y^k)$$

$$= b_{ij}y^ic_\ell^{jk}y^\ell$$

i.e.,

$$A_h = b_{ij} c^{jk}_{\ell} y^i y^{\ell} \frac{\partial}{\partial y^k}$$

The differential equations for the orbit curves of A_h are then:

$$\frac{dy^k}{dt} = b_{ij} c^{jk}_{\ell} y^i y^{\ell} \tag{6.5}$$

Let $t \to \theta(t)$ be the curve in $\mathcal{G}^d$ such that

$$y^k(t) \equiv y^k(\theta(t))$$

solves (6.5). Then,

$$\begin{aligned} y^k\left(\frac{d\theta}{dt}\right) &= b_{ij} c^{jk}_{\ell} y^i(\theta(t)) y^{\ell}(\theta(t)) \\ &= b_{ij} \theta(t)([X^j, X^k] y^i(\theta(t)) \\ &= \theta(t)([b_{ij} X^j y^i(\theta(t)), X^k]) \\ &= -\rho^d(\underset{\sim}{b}(\theta(t))(\theta(t))(X^k)) \end{aligned}$$

where:

$$\rho^d(X)(\theta)(Y) = -\theta([X,Y]) \quad , \tag{6.6}$$

for $\theta \in \mathcal{G}^d$; $X, Y \in \mathcal{G}$

i.e., $X \to \rho^d(X)$, a representation by $\mathcal{G}$ in $L(\mathcal{G}^d)$, is the dual of the adjoint representation of $\mathcal{G}$. b is given as a bilinear form in $\mathcal{G}^d$.

$$\underset{\sim}{b}: \mathcal{G}^d \to \mathcal{G}$$

is the linear map such that

$$\underset{\sim}{b}(\theta)(\theta') = b(\theta,\theta') \tag{6.7}$$

for $\theta,\theta' \in \mathcal{G}^d$

Thus, (6.6) can be rewritten in basis-free terms as:

$$\boxed{\frac{d\theta}{dt} = -\rho^d(\underset{\sim}{b}(\theta(t))(\theta(t))} \tag{6.8}$$

Remark. Formula (6.8) is a basic formula. It suggests a generalization--which might be significant and useful in a variety of contexts--of the Euler-Lax equation of inverse scattering theory. Namely, take an arbitrary Lie algebra $\mathcal{G}$, a vector space V, a linear representation $\rho: \mathcal{G} \to L(V)$, and a map $\underset{\sim}{b}: V \to \mathcal{G}$. For the equation:

$$\frac{dv}{dt} = \rho(\underset{\sim}{b}(v))(v) \tag{6.9}$$

This situation is enormously suggestive of "feedback" in systems theory. Indeed, there are many links between this subject and systems theory, which I intend to develop in later publications. (I am indebted to Roger Brockett for the first remark of this type, which set my research off in this direction.)

We can sum up as follows.

Theorem 6.1. Let h be a quadratic function of $\mathcal{G}^d$. Then A_h, the vector field associated to h via the standard cosymplectic structure on the dual space to a Lie algebra, is tangent to the orbits of $Ad^d\ G$ acting on $\mathcal{G}^d$. This vector field, restricted to each orbit, is an infinitesimal automorphism of the Kostant symplectic structure on the orbit.

Notice that Theorem 5.1 is now a corollary of Theorem 6.1. For, when $\mathcal{G}$ is reductive, there is a linear automorphism $\underset{\sim}{B}: \mathcal{G} \to \mathcal{G}^d$ (defined by a non-degenerate symmetric, bilinear form $B: \mathcal{G} \times \mathcal{G} \to R$ which is invariant under the adjoint representation) which intertwines the action of $Ad\,\mathcal{G}$ and $Ad^d(\mathcal{G}) \equiv \rho^d(\mathcal{G})$.

In certain circumstances (i.e., where the "generalized Toda lattice-Jacobi element-graded Lie associative algebra" situation described above holds), the orbits of A_h, when restricted to certain submanifolds of $\mathcal{G}$, are tangent to the orbits of a subgroup S of G. In the simplest situations (e.g., for the Toda lattice itself), S will be a <u>solvable</u> Lie group. This "solvability" property provides for the "solubility" of the equations of motion, in the sense of defining a set of "action-angle variables". Lie's general theorem on Function Groups--described in Part II--provides a (local) existence theorem for these objects. The point is to do it in an "effectively computable" way. In this case, it is provided by the group-theory. Certain recent authors have called this "complete integrability" of the equations of motion. This term is nonsense from the differential geometric point of view. It is a question, ultimately, of finding a complete solution of the Hamilton-Jacobi equation, which always *exists* (at least locally). No doubt, the vague (but inspiring!) nineteenth-century program of developing a "Galois theory" for differential equations is the appropriate context in which to think of these ideas.

BIBLIOGRAPHY

1. I.M. Gelfand and Levitan, "On the Determination of a Differential Equation from its Spectral Function", Amer. Math. Soc. Translation, Series 2, Vol. 1 (1955), 253-304.

2. V.A. Marchenko, "Reconstruction of Potential Energy in Terms of Stray Wave Physics", Dokl. Akad. Nauk. SSSR, 104 (1955), 695-698.

3. I. Kay and H. Moses, "The Determination of the Scattering Potential from the Spectral Measure Function, I", Nuova Cimento 2 (1955), 917-961.

4. I. Kay and H. Moses, "The Determination of the Scattering Potential from the Spectral Measure Function, II", Nuovo Cimento 3 (1956), 66084.

5. I. Kay and H. Moses, "The Determination of the Scattering Potential from the Spectral Measure Function, III", Nuovo Cimento 3 (1956), 276-304.

6. K. Case and M. Kac, "A Discrete Version of the Inverse Scattering Problem", J. Math. Phys. 14 (1973), 594-602.

7. H. Flaschka, "On the Toda Lattice, I & II", Phys. Rev. B9 (1974), 1924-1925 and Prog. Theor. Phys. 51 (1974), 703-716.

8. B. Kostant, Quantization and Unitary Representations, Lectures in Math., Vol. 170, Springer-Verlag.

THE DIFFERENTIAL-GEOMETRIC FOUNDATIONS OF THE INVERSE SCATTERING TECHNIQUE, IV

SOLVABILITY OF THE EQUATIONS OF MOTION AND MATHEMATICAL SYSTEMS THEORY

ABSTRACT

The results of Part III are used to illustrate and motivate development of certain general ideas, particularly in what sense a differential equation is "solvable". (This is the concept that is called--wrongly--"complete integrability" in the nonlinear wave literature.) Certain classical ideas about a possible "Galois Theory" of differential equations, developed around the turn of the century by Lie, Picard, Vessiot and Drach seem relevant here. There is also--as suggested by Roger Brockett--extensive contact with Mathematical Systems Theory, which will be briefly developed.

1. INTRODUCTION

Let G be a Lie group, V a vector space, $\rho: G \to L(V)$ a linear representation of G. Let $\mathcal{G}$ be the Lie algebra of G. ρ defines a linear representation, which we also denote by ρ, of $\mathcal{G} \to L(V)$.

Let $\alpha: V \to \mathcal{G}$ be a linear map. Form the differential equations:

$$\frac{dv}{dt} = \rho(\alpha(v))(v) \quad . \tag{1.1}$$

The right hand side of (1.1) defines a vector field X on the vector space V, whose orbit curves are the solutions of (1.1).

As we have seen, many equations that are intensely interesting for mathematical physics (and much else besides) can be obtained by suitably specializing (1.1). If $t \to v(t)$ is a solution of (1.1), then it lies on the orbit $\rho(G)(v(0))$.

However, we have seen examples (e.g., Toda lattice in a general algebraic version developed in Part III) where additional conditions are satisfied. There is a submanifold $N \subset V$ such that X is tangent to N. (In the "generalized Toda lattice" situation developed in Part III, N consists of the Jacobi elements of $V \equiv$ a graded associative algebra.) Let X_N denote the vector field X restricted to N. Our problem is to derive as much information as possible about the orbits of X_N.

Here is what seems to happen in the situation considered in Part III. There is a subgroup S of G with the following property:

> The solution curves of (1.1) which lie on N are contained in orbits of S.

This provides further information about the orbits of X_N and, in favorable circumstances, seems to lead to a complete determination of the solutions.

We shall now study certain general features of this situation.

I am indebted to Roger Brockett, whose suggestion to develop the relation between the Toda Lattice and Systems Theory was the catalyst for this work.

2. SOLVABILITY OF ORBIT CURVES OF VECTOR FIELDS ON MANIFOLDS

Let N be a manifold, and let X be a vector field on the manifold. Using the ordinary differential equation existence theory, we know that (modulo certain global complications, i.e., a finite "escape time") X generates a one-parameter group of diffeomorphisms of N. Of course, "Lie theory" in its original form involved this correspondence between one-parameter groups and vector fields.

Now, for purposes of applications, one would like to go beyond this pure existence theorem and determine the degree of "difficulty" of determining these orbit curves. In the 19-th century they thought of this as a sort of Galois theory for differential equations. The best description of the principles involved that I know of is Vessiot's "Encyclopédie" article [1]. They often "dualized" the problem--instead of the orbit curves of X, find the "constants of motion", i.e., the real-valued functions f on N which are constant along the orbit curves of X, i.e., those which satisfy:

$$X(f) = 0 \quad , \tag{2.1}$$

where one interprets the vector field as a first order linear differential operator, as usual in differential geometry. Now, Vessiot's idea seems to be to treat (2.1), a first order, linear partial differential equation, along the same lines as he earlier treated the linear ordinary differential equations, a subject that we now call the "Picard-Vessiot theory", and that has been absorbed into the standard "differential algebra".

One can replace (2.1) with the more general equation for f:

$$X(f) = Ff \quad , \tag{2.2}$$

where F is some function on N. We say that a function f satisfying (2.2) is a <u>relative constant of motion</u>. If $F = 0$, it, of course, reduces to (2.1). Otherwise,

$$X(\log f) = F \quad ,$$

i.e., if $t \to p(t)$ is an orbit curve of X,

$$\frac{d}{dt} \log(f)(p(t)) = F(p(t)) \quad ,$$

i.e.,

$$\log g(p(t)) - \log f(p(0)) = \int_0^t F(p(s))\, ds$$

or

$$f(p(t)) = f(p(0))\, e^{\int_0^t F(p(s))\, ds} \tag{2.3}$$

An even more general situation can be described as follows: Let V be a vector space, L a Lie group of automorphisms of V. We can consider maps $f: N \to N$ which satisfy

$$f(p(t)) = \ell(t) f(p(0)) \tag{2.4}$$

where $t \to \ell(t)$ is a curve in L.

My ultimate goal is to see how the functions satisfying (2.3) or (2.4) can be generated by group-theoretic principles for the vector fields defined by the generalized Toda lattice equations. I believe that, ultimately, there is some sort of Galois theory of the sort described by Vessiot involved here. As a simple, if trivial, illustration, we now consider the following example.

3. A "GALOIS THEORY" FOR LINEAR, CONSTANT COEFFICIENT VECTOR FIELDS

Suppose now that N is R^n. Denote its points by x. Suppose that:

$$X(x) = Ax \quad ,$$

where A is an $n \times n$ matrix. We can ask for linear functions

$$f(x) = y \cdot x$$

$$\text{for fixed } y \in R^n$$

which satisfy

$$X(f) = \lambda f \quad , \tag{3.1}$$

with λ constant. The condition for this is that

$$Ay = \lambda y \quad ,$$

i.e., that y is an eigenvector for A of eigenvalue λ.

In order to make a "Galois theory" out of this, we must suppose that the vector field is "known", and indicate some algorithms for assigning a degree of "complexity" to the problem of finding f. In this case, because of the purely algebraic nature of the situation, we can rely on the well understood, traditional case of "Galois theory", the solvability of algebraic equations in one variable "by radicals" [2,3]. The characteristic equation

$$\det(A-\lambda) = 0$$

for eigenvalues is, of course, an algebraic equation of degree n. Let K be the subfield of R generated by the rational numbers Q and the coefficients of the matrix A. Let $\lambda_1, \ldots, \lambda_n$ denote the eigenvalues, det K' be the subfield of $\mathbb{C}$ generated by K and the $\lambda_1, \ldots, \lambda_n$. The <u>Galois group</u>

of f mod K, i.e., the group of automorphisms of the field K' which leaves each element of K fixed, is, in a sense, the "measure of complexity" of the problem. For example, if it is a solvable group, the functions f which satisfy (3.1) can be constructed from K by "Liouville extensions", i.e., a rational formula involing K and m-th roots of elements of K. If the Galois group is not solvable, such an algorithm constructing f's is not possible.

Let us now turn from this strictly algebraic approach--which I hope to pursue further in later publications--in order to consider material more directly relevant to the main theme of these paper.

4. A CONSTRUCTION OF RELATIVE CONSTANTS OF MOTION

Let M be a manifold, X a vector field on M, N a submanifold of M. Let G be a Lie group acting on M. Suppose that the following conditions are satisfied:

a) X is tangent to N

b) Each orbit of X starting on N lies on a single orbit of the group G.

Let us now translate these geometric assumptions (which are satisfied by the "Toda lattice" and the "generalized Toda lattices defined by graded associative algebras" which are defined in Part III) into a more analytical form. Let $\mathcal{G}$ be the Lie algebra of G. The geometric action of G on M defines an infinitesimal action of $\mathcal{G}$ as a Lie algebra of vector fields on M. Let

$$(Y^i) \quad , \qquad 1 \leq i,j \leq n \quad ,$$

be a basis for $\mathcal{G}$. Then our assumptions mean that there are relations of the following form:

$$X = f_i Y^i + Z \tag{4.1}$$

where the f_i are functions on M, and Z is a vector field on M such that

$$Z(p) = 0 \tag{4.2}$$

$$\text{for } p \in N \quad .$$

Now, let V be a real vector space, and let

$$\rho\colon \mathcal{G} \to L(V)$$

be a linear representation of $\mathcal{G}$ by linear maps on V. Set up the following auxiliary differential equation system, to be solved for a map

$$h\colon M \to V \quad .$$

$$Y(h) = -\rho(Y)(h) \tag{4.3}$$

$$\text{for all } Y \in \mathcal{G} \quad .$$

We see now that--as a system of first order linear partial

differential equations--it is "completely integrable" in the classical Mayer-Frobenius sense. This means that the compatibility relations deduced from (4.3) by differentiating both sides are identically satisfied. This can be seen by applying another vector field $Y' \in \mathcal{G}$ to both sides of (4.3):

$$Y'(Y(h)) = -Y'(\rho(Y)(h))$$

$$= -\rho(Y)(Y'(h))$$

$$= \rho(Y)\rho(Y')(h) \tag{4.4}$$

= , working on the left hand side,

$$[Y',Y](h) \neq Y(Y'(h))$$

= , using (4.3) again,

$$-\rho([Y',Y]/h) - Y(\rho(Y')(h))$$

$$= -\rho([Y',Y])(h) - \rho(Y')(Y(h))$$

$$= -\rho([Y',Y])(h) + \rho(Y')\rho(Y)(h)$$

$$= \rho(Y)\rho(Y')(h) \tag{4.5}$$

Equality of (4.4) and (4.5) (which is to be expected if h satisfies (4.3)), and which is now satisfied identically, is the criteria of complete integrability.

Remarks. a) This classical argument can be replaced with a more "global" version. Construct a foliation on

$$M \times V$$

whose leaves are precisely the graphs of the maps $M \to V$ which satisfy (4.3).

b) The Mayer-Frobenius existence theorem would break down at points where the rank of the vector fields (Y^i) was not maximal, i.e., at the singular orbits of the action of G on M. Thus, as usual in the classical literature, the existence for h's satisfying (4.3) is only proven "generically".

Now, we can use the h's satisfying (4.3) and relations (4.2)-(4.3) to say something significant about the orbits of X on the submanifold N. Namely,

$$\boxed{X(h) \text{ restricted to } N = f_i \rho(Y^i)(h)} \tag{4.6}$$

Geometrically, this can be interpreted as follows. Suppose that

$$t \to x(t)$$

is a curve in N which is an orbit curve of X. Then,

$$\boxed{\frac{d}{dt} h(x(t)) = \beta(t)(h(x(t)))} \tag{4.7}$$

where, for each t, $\beta(t)$ is a linear map: $V \to V$ which belongs to the Lie algebra $\rho(\mathcal{G})$ of linear maps on V.

If we now specialize G to a solvable group and choose the representation ρ in an appropriate way, h restricted to N turns out to be a relative constant of motion.

5. RELATIVE CONSTANTS OF MOTION DETERMINED BY SOLVABLE LIE GROUPS

Keep the notation of Section 4, but specialize G to be a solvable Lie group. Now, solvable Lie groups always admit one-dimensional linear representation. Choose ρ to be one. Then, h is a map

$$M \to R \quad .$$

$\rho(Y)$ is now a real number, for each $Y \in \mathcal{G}$. (4.6) takes the form

$$\boxed{X(h) \text{ restricted to } N = Fh} \tag{5.1}$$

with $F = f_i \rho(Y^i)$, i.e., h is a relative constant of motion for the orbits of X restricted to N.

6. CONSTRUCTION OF RELATIVE CONSTANTS OF MOTION BY MEANS OF SYSTEMS THEORY. MOSER'S APPROACH TO THE TODA LATTICE

Let us recapitulate notation. M is a manifold with N a submanifold. X is a vector field on M which is tangent to N. G is a Lie group acting on M. The orbits of X_N (X restricted to N) lie on orbits of G.

Now, let V, W, W' be complex vector spaces (at the moment assume that they are finite dimensional, but keep in mind that much of the broad outline of the ideas carries over to the infinite dimensional situations, although, of course, the technicalities become harder). A <u>linear system</u> [4,5] with V as <u>state space</u>, W as <u>input-space</u>, W' as <u>output space</u>, is a triple

$$(A,B,C) \quad ,$$

with

$$A \in L(V,V) \equiv \text{space of linear maps } V \to V$$

$$B \in L(W,V) \equiv \text{space of linear maps } V \to W$$

$$C \in L(V,W') \equiv \text{space of linear maps } V \to W' \quad .$$

Such a system is denoted as $\sigma = (A,B,C)$. The collection of them is denoted as Σ. (Then, $\Sigma = L(V,V) \times L(W,V) \times L(V,W')$.)

Let $GL(V)$ denote the group of invertible linear maps $V \to V$. Let it act on Σ as follows

$$g(A,B,C) = (gAg^{-1},B,C) \quad . \tag{6.1}$$

<u>Remark</u>. This is <u>not</u> the action which is customary in systems theory! The action $g(A,B,C) = (gAg^{-1}, gB, Cg^{-1})$ is usually considered; it is called <u>systems equivalence</u>.

<u>Definition</u>. A <u>system representation</u> of (M,N,X,G) is defined as a map $\phi\colon M \to L(V)$ and a linear representation $\rho\colon G \to GL(V)$ such that:

$$\phi(gp) = \rho(g)\phi(p)\rho(g^{-1}) \tag{6.2}$$

For $g \in G$, $p \in M$. (In words, ϕ is an <u>intertwining map</u> for the action of G.)

Let us suppose that such a system representation is given. Now, the chief object of interest in systems theory are the <u>input-output relations</u> [4,5] determined by $\sigma = (A,B,C) \in \Sigma$. They are determined by the <u>transfer function</u>

$$T(s) = C(s-A)^{-1}B \tag{6.3}$$

which is a meromorphic function of the complex variable, s, with values in $L(W,W')$. Our goal is to find information about the orbits X in terms of the action of $\rho(G)$ on the transfer functions. (In a disguised form, this is what happens in Moser's ingenious treatment [6] of the Toda Lattice.) To do this, let us <u>suppose</u> that B and C, the "input" and "output" maps, satisfy the following condition:

There exist representations

$$\rho'\colon G \to L(W), \quad \rho''\colon G \to L(W'')$$

such that:

$$\rho(g)B = B\rho'(g) \tag{6.4}$$

$$C\rho(g) = \rho''(g)C \quad . \tag{6.5}$$

(In words, ρ' and ρ'' are representations for which B and C are intertwining maps.)

With these special assumption, we can determine the action of G on transfer functions:

If $T = C(s-A)^{-1}B$, $T' = C(s-A')^{-1}B$, with $A' = \rho(g)A\rho(g)^{-1}$, then

$$T' = C\rho(g)(s-A)^{-1}\rho(g^{-1})B$$

$$= \text{, using (6.4)-(6.5),}$$

$$\rho''(g)T\rho'(g^{-1}) \quad , \tag{6.6}$$

i.e., T transforms as an <u>invariant</u> under the representation $\rho'' \otimes \rho'^{d}$.

In particular, if

$$t \to p(t) \in N$$

is an orbit curve of X, and if

$$A(t) = \phi(p(t)) \quad ,$$

$$T_t(s) = C(s-A(t))^{-1}A \quad ,$$

Then there is a curve $t \to g(t)$ in G such that

$$T_t(s) = \rho''(g(t))T_0(s)\rho'(g(t)^{-1}) \quad . \tag{6.7}$$

We can use this formula to construct relative invariants for the orbits of X_N following Moser [6], i.e., using the <u>partial fraction decomposition</u> of $T_t(s)$

$$T(s) = \frac{r_1}{(s-\lambda_1)} + \frac{r_{12}}{(s-\lambda_1)^2} + \cdots + \frac{r_2}{(s-\lambda_2)} + \cdots \\ + \cdots \quad , \tag{6.8}$$

where $\lambda_1,\ldots,\lambda_n$ are the eigenvalues of A, $r_1,r_2,\ldots$ (which are elements of $L(W,W')$) are the <u>residues</u>. Now, the eigenvalues of A are invariant under the action of $\rho(G)$. Hence, the residues $r_1,r_{12},\ldots,r_2,\ldots$ also transform like (6.7). They determine the map

$$M \to L(W,W') \quad .$$

On each orbit curve of X_N, the values $r_1(t),\ldots$ of the residues also transform like (6.6).

In particular, let us suppose that the group σ is <u>solvable</u>. (Recall that this is the key group-theoretic property of the Toda lattice situation, and of the Gelfand-Levitan algorithms.) In this case, $\rho(G)$ acting in V has common eigenvectors (Lie's theorem). Hence, we can <u>choose</u>

$$W = W' = \mathbb{C} \quad ,$$

and the residue functions now become complex-valued functions on M, which were what we earlier called <u>relative constants of motion</u> for the vector field X_N.

In order to complete the program of developing the material in Moser's paper [6] in a general differential-geometric setting, it only remains to determine the "Poisson bracket" properties of these residue functions when M has a cosymplectic structure. This will be the subject of a later paper.

BIBLIOGRAPHY

1. E. Vessiot, "Méthodes d'integration élémentaire. Etude des équations différentielles ordinaires au point de vue formel", Encyclopédie des sciences mathematiques (1910), Tome II, Vol. 3, 58-170.

2. I. Stewart, Galois Theory, Chapman and Hall, London, 1973.

3. R. Hermann, The Geometry of Nonlinear Differential Equations, Bäcklund Transformations and Solitons, Part A (Interdisciplinary Mathematics, Volume 12), Math Sci Press, Brookline, Mass., 1976.

4. R. Brockett, Finite Dimensional Linear Systems, Wiley, N.Y., 1970.

5. R. Hermann, Linear Systems Theory and Introductory Algebraic Geometry (Interdisciplinary Mathematics, Volume 8), Math Sci Press, Brookline, Mass., 1974.

6. J. Moser, Finitely Many Mass Points on the Line Under the Influence of an Exponential Potential--An Integrable System, Lecture Notes in Physics, Volume 38, Springer-Verlag, 1975.

THE DIFFERENTIAL-GEOMETRIC FOUNDATION OF THE INVERSE SCATTERING TECHNIQUE, V

THE IMPULSE RESPONSE OF LINEAR SYSTEMS THEORY AND THE TODA LATTICE

1. INTRODUCTION

A good deal of the motivation of this series of papers is the desire to understand the ramifications into differential geometry and systems theory of Moser's solution [1] of the Toda Lattice equations. P. Krishnaprasad has remarked [private communication] that Moser's solution has a good deal to do with what systems theorists call the "shift" operation on linear systems [2-4]. In this paper I shall investigate this relation in more detail, and show that it is clearly related to the material in the earlier work of this series in graded associative algebras.

Let A denote an $n \times n$ real Jacobi matrix. Let B, C be elements of R^n, and form:

$$W(\tau) = C' e^{\tau A} B \quad . \tag{1.1}$$

W is called the <u>impulse response</u> (or the <u>weighting pattern</u>, C' is the transpose of C). Let $\lambda_1,\ldots,\lambda_k$, $k = 1,\ldots,n$, be the eigenvalues of A. Suppose that $r_1,\ldots,r_n$ are the numbers such that

$$W(\tau) = \sum_{k=1}^{n} r_k^2 e^{\lambda_k \tau} \tag{1.2}$$

Now, let

$$t \to A(t)$$

be the orbits of the flow in the Jacobian matrices generated by the Toda Lattice equations. Set:

$$\begin{aligned} W(t,\tau) &= C' e^{\tau A(t)} B \\ &= \sum_k r_k(t)^2 e^{\lambda_k(t)\tau} \end{aligned} \tag{1.3}$$

According to the material on page 479 of [1], the following formulas hold:

$$\lambda_k(t) = \lambda_k(0) \tag{1.4}$$

$$\frac{dr_k}{dt} = -\left(\lambda_k - \sum_j \lambda_j r_j^2\right) r_k \tag{1.5}$$

Hence,

$$\frac{\partial W}{\partial t}(t,\tau) = \sum_k 2\frac{dr_k}{dt} r_k e^{\lambda_k \tau}$$

$$= -2 \sum_k \left(\lambda_k - \sum_j \lambda_j r_j^2\right) r_k^2 e^{\lambda_k \tau}$$

$$= -2\frac{\partial W}{\partial \tau}(t,\tau) + 2\frac{\partial W}{\partial \tau}(t,0)W(t,\tau) \qquad (1.6)$$

This is a differential equation for the Impulse Response that replaces the equations (1.4)-(1.5) for the eigenvalues and residues. It is obviously much better suited for general purposes! We will now examine the conditions it places on the system (A,B,C).

2. ANALYSIS OF THE IMPULSE RESPONSE EQUATIONS IN TERMS OF GRADED ASSOCIATIVE ALGEBRAS

Suppose that

$$\frac{dA}{dt} = [\alpha, A] \quad , \qquad (2.1)$$

where $t \to \alpha(t)$ is a family of matrices. Then,

$$\begin{aligned} \frac{\partial W}{\partial \tau}(t,\tau) &= C'[\alpha, \exp(\tau A(t))]B \\ &= (\alpha' C)' \exp(\tau A)B - C' \exp(\tau A)(\alpha B) \end{aligned} \qquad (2.2)$$

$$\frac{\partial W}{\partial \tau}(t,\tau) = C' \exp(\tau A)AB \qquad (2.3)$$

$$\frac{\partial W}{\partial \tau}(t,0) = C'AB \tag{2.4}$$

Following [1], let us now make A and α more explicit:

$$A = \begin{pmatrix} b_1 a_1 & & & 0 \\ a_1 b_2 & & & \\ & \ddots & & \\ 0 & & & b_{n-1} a_{n-1} \\ & & & a_{n-1} b_n \end{pmatrix}$$

$$\alpha = \begin{pmatrix} 0 & a_1 & \\ -a_1 & 0 & \\ & & \ddots \end{pmatrix}$$

Set:

$$A_0 = \begin{pmatrix} b_1 & & 0 \\ & \ddots & \\ 0 & & b_n \end{pmatrix}$$

$$A_1 = \begin{pmatrix} 0 & a_1 & & 0 \\ & \ddots & \ddots & \\ & & \ddots & a_{n-1} \\ 0 & & & 0 \end{pmatrix}$$

$$A_{-1} = \begin{pmatrix} 0 & & & 0 \\ & \ddots & & \\ a_1 & & & \\ & \ddots & & 0 \\ 0 & & a_{n-1} & \end{pmatrix}$$

Then,

$$A = A_{-1} + A_0 + A_1 \qquad (2.5)$$

$$\alpha = A_1 - A_{-1} \qquad (2.6)$$

This defines a graded associate algebraic structure $\mathscr{A}$ on the algebra of $n \times n$ real matrices.

$$\mathscr{A} = \cdots \mathscr{A}_{-1} \oplus \mathscr{A}_0 \oplus \mathscr{A}_1 \oplus \cdots$$

Let us now suppose that B and C satisfy the following conditions:

$$B = C \qquad (2.7)$$

$$A_{-1}(B) = 0 \qquad (2.8)$$

$$A_0(B) = bB \qquad (2.9)$$

for $b \in R$, i.e., B is a common eigenvector for $\mathscr{A}_0$.

$$B'B = 1 \quad . \qquad (2.10)$$

Let us now introduce these assumptions into (2.2)-(2.4):

$$\frac{\partial W}{\partial \tau}(t,\tau) = -(\alpha C)' \exp(\tau A)B - C' \exp(\tau A)(A,B)$$

$$= -(A,B)' \exp(\tau A)B - B' \exp(\tau A)(A,B)$$

$$2\,\frac{\partial W}{\partial \tau}(t,\tau) = (AC)'\exp(\tau A)B + C'\exp(\tau A)(AB)$$

$$= (A,B)'\exp(\tau A)B + C'\exp(\tau A)(A,B) + 2bB'\exp(\tau A)B$$

$$= (A,B)'\exp(\tau A)B + B'\exp(\tau A)(A,B) + 2bW(t,\tau)$$

Hence,

$$\frac{\partial W}{\partial \tau}(t,\tau) + 2\,\frac{\partial W}{\partial \tau}(t,\tau) = 2bW(t,\tau) \tag{2.11}$$

Finally,

$$\frac{\partial W}{\partial \tau}(t,0) = B'AB = B'A_0B = b \tag{2.12}$$

(2.11) and (2.12) together prove formula (1.6), under the assumptions (2.7)-(2.10). It would be interesting to know all possible solutions of Equation (1.7), especially for more general choices of graded associative algebras!

3. FINAL REMARKS

With Moser [1], we can consider the map

$$A \to (\lambda_1,\ldots,\lambda_n;\ r_1,\ldots,r_n)$$

of the set of Jacobi matrices into R^n. Moser shows that

Equation (1.5) (hence a fortiori Equation (1.7)), leads to the following transformation property of these functions under the Toda Lattice flow:

$$\lambda_k(t) = \lambda_k(0) \tag{3.1}$$

$$r_k^2(t) = \frac{r_k^2(0)e^{-2\lambda_k t}}{\Sigma r_j^2(0)e^{-2\lambda_j t}} \tag{3.2}$$

Formulas (3.1)-(3.2) can most conveniently be interpreted in terms of classical mechanics:

λ_k are the "action variables" of the system

r_k are the "angle variables"

What has been done is essentially to find explicitly (or "algebraically") a complete solution of the Hamilton-Jacobi equation associated with the mechanics problem.

I would be very interested to know if Equation (1.6) is satisfied for any <u>infinite dimensional</u> system, e.g., the isospectral flow that the Korteweg-de Vries equation induces on the linear Schrödinger operators.

BIBLIOGRAPHY

1. J. Moser, Finitely Many Mass Points on the Line Under the Influence of an Exponential Potential--An Integrable System, *Lecture Notes in Physics*, Vol. 38 (ed., J. Moser), Springer-Verlag, 1975.

2. R. Brockett, *Finite Dimensional Linear Systems*, J. Wiley, New York, 1970.

3. R. Brockett and P. Krishnaprased, Scaling Rational Functions and Linear System Identification, Harvard Univ. preprint, 1977.

4. R. Hermann, *Linear Systems Theory and Introductory Algebraic Geometry* (Interdisciplinary Mathematics, Vol. 8), Math Sci Press, Brookline, Mass., 1974.

Chapter 1

BÄCKLUND TRANSFORMATIONS IN THE SENSE OF ESTABROOK AND WAHLQUIST

1. INTRODUCTION

As we have seen, the concept of "Bäcklund transformation" is rather flexible and loose in both the classical and modern literature. In order to tie it down better (so as to better understand!) I have developed (in Vol. XII) an approach which emphasizes the analogy with "correspondence" in algebraic geometry.

Estabrook and Wahlquist suggested--in their landmark paper [1]--another definition which arose out of their work on the Bäcklund transformation for the Korteweg-de Vries equation [2]. As Hedly Morris has shown [3] this definition is extremely interesting in its own right. I will now describe a few of its features. In fact, it is just a special type among those which are generalization of "correspondence".

2. PROLONGATIONS AND BÄCKLUND TRANSFORMATIONS

Let $\pi: Y \to X$ be a submersion map between manifolds. (For our purposes, it would suffice to think of Y as a product of X and a fiber manifold, with π the Cartesian projection.) Let ED be an exterior differential system in X, i.e., a collection of differential forms which forms an ideal in the Grassman algebra sense and also is closed under exterior derivative.

A prolongation (in the sense of Estabrook and Wahlquist) is a collection P of forms on Y such that:

$$dP \subset \text{Grassman algebra ideal generated by } P \text{ and } \pi^*(ED). \tag{2.1}$$

A Bäcklund transformation in their sense is now a diffeomorphism

$$\phi: Y \to Y$$

(not necessarily fiber preserving!) such that

$$\phi^*(\pi^*(ED)) \subset \pi^*(ED) + \text{Grassman algebra ideal generated by } P. \tag{2.2}$$

Now, in order to give this the name "Bäcklund transformation", we must see that it has something to do with a "transformation" on the solution submanifolds of ED. This we now do.

3. ACTION ON THE SOLUTION SUBMANIFOLDS OF ED. RELATIONS WITH THE GENERAL CONCEPTS OF "BÄCKLUND TRANSFORMATION"

Let $\alpha \subset X$ be a solution submanifold of ED. Set:

Γ_α = set of maps $\gamma: \alpha \to Y$ such that:

$$\pi\gamma(x) \quad = \quad x \tag{3.1}$$

for all $x \varepsilon \alpha$

$$\gamma^*(P) \quad = \quad 0 \tag{3.2}$$

(In the relevant examples, Γ_α might consist of the solutions of the "inverse scattering problem" associated with the "non-linear wave equation" defined by ED.)

Now, let

$$\phi\colon Y \to Y$$

be a map satisfying (2.2), i.e., defining a Bäcklund transformation in the Estabrook-Wahlquist sense

For each pair (α,γ), with $\alpha \subset X$ a solution submanifold of ED, $\gamma \varepsilon \Gamma_\alpha$, set:

$$\phi(\alpha,\gamma) \quad = \quad \pi\phi\gamma\alpha \tag{3.3}$$

Consider $\phi(\alpha,\gamma)$, as defined by (3.3), as a submanifold of X. It is defined via the following commutative diagram of mappings:

$$\begin{array}{ccc} & Y & \xrightarrow{\ \phi\ } & Y \\ & \uparrow{\scriptstyle \gamma} & & \downarrow{\scriptstyle \pi} \\ \alpha\colon S \to & X & \xrightarrow[\ \phi(\alpha,\gamma)\ ]{} & X \end{array}$$

<u>Theorem 3.1</u>. $\phi(\alpha,\gamma)$ is a solution submanifold of ED.

Proof.

$$\phi(\alpha,\gamma)^*(ED) = (\pi\phi\gamma\alpha)^*(ED)$$
$$= \alpha^*\gamma^*\phi^*\pi^*(ED) \quad ,$$

which is contained in, by (2.2),

$$\alpha^*\gamma^*(\pi^*(ED) + F^{(\ )}(Y) \wedge P) \subset \alpha^*((\gamma\pi)^*(ED) + 0) \quad ,$$

by using (3.2)

$$\subset \alpha^*(ED) = \underset{\equiv}{0}$$

There is an interesting abstract structure involved here. Let $\mathscr{S}$ denote the solution submanifolds of ED, i.e., the "space" of solutions of the given non-linear wave equation. Let $\mathscr{S}_P$ = space of pairs

(α,γ) ,

with $\alpha \in \mathscr{S}$, $\gamma \in \Gamma_\alpha$

There is a map $\mathscr{S}_P \to \mathscr{S}$, assigning α to (α,γ), which in a sense defines $\mathscr{S}_P$ as a "fiber space" over $\mathscr{S}$. (In the cases covered by the Estabrook-Wahlquist ideas, the fiber of this map is finite dimensional. In the "inverse scattering situations" it is a <u>vector space</u>, so the "prolongation" construction gives essentially a <u>vector bundle</u> whose base is $\mathscr{S}$.)

The map

$$(\alpha,\gamma) \to \phi(\alpha,\gamma)$$

now defines a map

$$\mathcal{S}_P \to \mathcal{S} \tag{3.4}$$

Suppose we exhibit $\mathcal{S}_P$ as the product

$$\mathcal{S} \times Z \quad ,$$

where Z is a space. Then, the map (3.4) defines a map

$$\mathcal{S} \times Z \to \mathcal{S} \qquad .$$

This gives a new set-theoretic meaning to a "Bäcklund transformation" as a "symmetry" of a system of differential equations. A "symmetry" in the usual sense (e.g., as made most precise by Lie and Cartan) is a mapping $\mathcal{S} \to \mathcal{S}$. We see in what sense "Bäcklund transformations" might be considered as a generalization.

Following the analogy with "correspondences" in the sense of algebraic geometry (as described in Part A, for example), a "Bäcklund transformation" <u>generally</u> should mean a "relation" between $\mathcal{S}$, i.e., a subset $\mathcal{B}$ of $\mathcal{S} \times \mathcal{S}$. We can see how the Estabrook-Wahlquist idea gives a realization of this: Let

$$\mathcal{B} = \text{set of } (\alpha,\alpha') \in \mathcal{S} \times \mathcal{S} \text{ such that there equals } \gamma \in \Gamma_\alpha \text{ with } \alpha' = (\alpha,\gamma) \quad .$$

We can also describe "Bäcklund transformations" within a "pure" context of Exterior Differential Systems. Here is a suitable definition:

> Let ED, ED' be exterior differential systems on manifolds X,X'. A <u>Bäcklund system interrelating them</u> is a quadruple
>
> $$(X'', ED'', \pi: X'' \to X, \quad \pi': X'' \to X) \quad ,$$
>
> with ED" an exterior system on X" such that
>
> $$\pi'^*(ED) \subset ED''$$
>
> $$\pi'^*(ED') \subset ED''$$

Let $\mathcal{S}$, $\mathcal{S}'$ be the set of solution manifolds of ED, ED'. Such a Bäcklund system enables us to define a subset

$$\mathcal{B} \subset \mathcal{S} \times \mathcal{S}' \quad .$$

Namely, $\mathcal{B}$ consists of the pairs

$$(\alpha,\alpha') \in \mathcal{S} \times \mathcal{S}'$$

such that there exists a solution submanifold α'' of ED" such that

$$\pi(\alpha'') = \alpha \; ; \quad \pi'(\alpha'') = \alpha' \quad .$$

In our article "Differential Geometric Prolongations and Bäcklund Transformations", which will appear in the Proceedings of the 1976 Ames Conference [3], Estabrook, Wahlquist and I have shown how the prototypical Sine-Gordon Bäcklund appears in this context

That calculation may now be reinterpreted in the framework described in this section.

4. THE SINE-GORDON BÄCKLUND AS AN EXAMPLE OF AN ESTABROOK-WAHLQUIST BÄCKLUND TRANSFORMATION

Now specialize the general framework of Section 3. Let

X	=	space of variables (u, u_x, u_t, x, t)
θ_1	=	$du - u_x dx - u_t dt$
θ_2	=	$du_t \wedge dt - du_x \wedge dt - 2(\sin u) dx \wedge dt$
ED	=	exterior differential system in X generated by θ_1, θ_2, $d\theta_1$, $d\theta_2$

The two-dimensional solution submanifolds of ED on which $dx \wedge dt \neq 0$ (which is what we shall mean by "solution submanifold") are then the maps

$$(x,t) \to (x, t, u(x,t), u_x(x,t), u_t(x,t))$$

such that:

$$u_x(x,t) = \frac{\partial u}{\partial x}(x,t)$$

$$u_t(x,t) = \frac{\partial u}{\partial t}(x,t)$$

$$\frac{\partial^2 u}{\partial x \partial t} = u_{xt} = \sin u$$

i.e.,

$$(x,t) \to u(x,t)$$

is a solution of the Sine-Gordon equation.

Now, introduce a "pseudopotential variable" y, a manifold

$$Y = \text{space of variables } (x,t,u,u_x,u_t,y) ,$$

with $\pi: Y \to X$ the Cartesian projection, with y the coordinate of the fiber. Set:

$$\eta = dy - \left(u_x + 2\lambda \sin\left(\frac{y+u}{2}\right)\right) dx - \left(\frac{2}{\lambda} \sin\left(\frac{y-u}{2}\right) - u_t\right) dt \quad (4.1)$$

One readily verifies that:

$d\eta$ lies in the Grassman algebra ideal generated by θ_1, θ_2, and η.	(4.2)

This is the "prolongation" property defined by Estabrook and Wahlquist-- y is the "pseudopotential" for the Sine-Gordon equation.

Now, define the map

$$\phi: Y \to Y$$

by the following formula:

$$\phi(x,t,u,u_x,u_t,y) = \left(x,\ t,\ u_x + 2\lambda \sin\left(\frac{y+u}{2}\right),\ -u_t + \frac{2}{\lambda}\left(\frac{y-u}{2}\right),\ y\right) \quad (4.3)$$

It can be proved (and the calculation may be found in the paper cited above) that ϕ defined in this way is a "Bäcklund transformation" in Estabrook and Wahlquist's sense.

5. THE BÄCKLUND TRANSFORMATION IN THE SENSE OF ESTABROOK AND WAHLQUIST AS AN AUTOMORPHISM IN THE INVERSE SCATTERING CONTEXT

In the work of those who approach "non-linear wave theory" via the Inverse Scattering Technique, there is another way of looking at Bäcklund transformations. Roughly, they try to find "automorphisms" of the _linear_ equations determining the Inverse Scattering. See the articles in "Bäcklund Transformations", the Springer Verlag Math Lecture Note volume edited by R. Muira, for many examples of this sort of thing.

Let

$$\pi \colon Y \to X$$

be a _vector bundle_ over a manifold X. Suppose a linear connection is given. It determines a Pfaffian system P on Y. The curvature of the connection determines an exterior differential system (generated by 2-forms) ED on X.

Let Z be a fixed two-dimensional manifold. Consider submanifold maps

$$\alpha \colon Z \to X \quad .$$

Let $\mathscr{S}$ be those submanifold maps over which the connection is _flat_, i.e., such that

$$\alpha^*(ED) = 0 \quad .$$

For $\alpha \in \mathscr{S}$, let Γ_α be the space of maps $\gamma: Z \to Y$ such that

$$\pi\gamma = \alpha \quad , \tag{5.1}$$

and

$$\gamma^*(P) = 0 \tag{5.2}$$

(In words, Γ_α consists of the cross-section maps of the pulled-back vector bundle

$$\alpha^{-1}(Y) \quad .)$$

Γ_α is a <u>vector space</u>.

(5.2) represents a set of (completely integrable, i.e., "Mayer" type--see Part A) differential equations for Γ_α. In particular, Γ_α is a <u>finite dimensional</u> vector space.

If Y is taken as the product $X \times V$, where V is a vector space (and this is the typical situation in the applications!) Γ_α can be considered as the subset

$$M(Z,V) \ (\equiv \text{space of maps } Z \to V)$$

which satisfy a system of <u>linear</u> differential equations

$$D_\alpha = 0$$

which <u>depend on</u> α. (This assignment of a set of <u>linear differential equations</u> which <u>become completely integrable</u> when α satisfies a given set of <u>non-linear differential equations</u>, is, of course, the main feature of the Inverse Scattering Technique.

Now, let

$$\phi: Y \to Y$$

be a map which defines a Bäcklund transformation in the sense of Estabrook and Wahlquist. As we have seen in the previous section, this defines a map

$$\overline{\phi}: \mathscr{S}_\alpha \to \mathscr{S}$$

Namely, assign to $\gamma \in \mathscr{S}_\alpha$, which is a map $Z \to Y$, the map

$$\pi\phi\gamma \quad .$$

Let $\Gamma_{\mathscr{S}}$ denote the vector bundle over $\mathscr{S}$ defined as follows.

$$\Gamma_{\mathscr{S}} = \{(\alpha,\gamma): \gamma \in \Gamma_\alpha\}$$

The vector bundle projection map: $\Gamma_{\mathscr{S}} \to \mathscr{S}$ is the map $(\alpha,\gamma) \to \alpha$.

$\phi: Y \to Y$ defines the map $\overline{\phi}: \Gamma_{\mathscr{S}} \to \mathscr{S}$

Consider the map

$$\gamma \to \phi\gamma$$

as a map of $\Gamma_{\mathscr{S}} \to \Gamma_{\mathscr{S}}$.

Let us make this more explicit in the traditional terms. Suppose

Z = space of variables (x,t)

X = space of variables (x,t,u), with $u \in R^n$.

An α is then a map $(x,t) \to u(x,t)$.

Let $Y = X \times V$, with V a real vector space. Introduce variables $u_x, u_t, u_{xx}, \ldots$ Suppose Γ consists of the maps

$$v\colon (x,t) \to v(x,t) \text{ of } R^2 \to V .$$

Let D_α be a linear differential equations of the form

$$\begin{aligned} v_x &= A(u,u_x,u_t,\ldots)(v) \\ v_t &= B(u,u_x,u_t,\ldots)(v) \end{aligned} \tag{5.3}$$

($A(\ )$, $B(\ )$ are maps which take values in the space of linear maps $V \to V$.)

ϕ now is to be defined by the following formula:

$$(v,u,u_x,u_t,\ldots) = \phi(u,u_x,\ldots\ v) , \tag{5.4}$$

The process now starts off with a function $u(x,t)$ (i.e., a submanifold map $\alpha\colon Z \to X$) such that the equations (5.3) are completely integrable. Denote this map by $\underline{u}$. Pick a solution $v(x,t)$ (denoted by $\underline{v}$) of (5.3). $\Gamma_{\mathscr{S}}$ then consists of the pairs

$$(\underline{u},\underline{v}) .$$

We now require that

$$\underline{v}' = \phi(\underline{u},\underline{u}_x,\ldots,\underline{v}) \tag{5.5}$$

be the solution of another equation of form (5.3), but with $\underline{u}$ replaced by another $\underline{u}'$, i.e.,

$$\underline{v}'_x = A(\underline{u}',\ldots)(\underline{v}')$$

$$\underline{v}'_t = B(\underline{u}',\ldots)(\underline{v}')$$

The "transformation"

$$(\underline{u},\underline{v}) \to \underline{u}'$$

is then the "Bäcklund transformation".

Let us now turn to the example of the Korteweg-de Vries equation in order to study this technique.

6. THE KORTEWEG-DE VRIES EQUATION FROM THE CONNECTION POINT OF VIEW

Start off with the "Schrödinger" (i.e., Sturm-Liouville) differential equation

$$v_{xx} = uv \quad . \tag{6.1}$$

Here, v and u are scalar functions of (x,t). Add to this an equation of the form:

$$v_t = u_1 v + u_2 v_x \quad , \tag{6.2}$$

where u_1, u_2 are unknown functions. Can we choose them so that the system (6.1)-(6.2) is <u>completely integrable</u>? To do this, differentiate (6.1) with respect to t, (6.2) with respect to x <u>twice</u>, equate, and use the equations themselves to reduce the resulting expression in terms of v, v_x:

$$v_{xxt} = u_t v + u(u_1 v + u_2 v_x) \tag{6.3}$$

$$\begin{aligned} v_{txx} &= (u_1 v + u_2 v_x)_{xx} \\ &= u_{1xx} v + 2u_{1x} v_x + u_1 v_{xx} + u_{2xx} v_x + 2u_{2x} v_{xx} + u_2 v_{xxx} \\ &= u_{1xx} v + 2u_{1x} v_x + u_1 (uv) + u_{2xx} v_x + 2u_{2x} (uv) + u_2 (uv)_x \\ &= (u_{1xx} + u_1 u + 2u_{2x} u + u_x u_2) v + (2u_{1x} + u_{2xx} + uu_2) v_x \end{aligned} \tag{6.4}$$

Equate the right hand sides of (6.3)-(6.4), then equate the coefficients of v_1 and v_x:

$$u_t + uu_1 = u_{1xx} + u_1 u + 2u_{2x} u + u_x u_2 \tag{6.5}$$

$$uu_2 = 2u_{1x} + u_{2xx} + uu_2 \tag{6.6}$$

(6.5) and (6.6) are then equivalent to

$$u_t = u_{1xx} + 2u_{2x} u + u_x u_2 \tag{6.7}$$

$$2u_{1x} + u_{2xx} = 0 \quad , \tag{6.8}$$

or

$$2u_1 = -u_{2x} \tag{6.9}$$

Insert (6.9) into (6.7):

$$u_t = -\frac{1}{2} u_{2xxx} + 2u_{2x} u + u_x u_2 \tag{6.10}$$

In order to solve (6.10), we can prescribe u_2 arbitrarily. For example, set

$$u_2 = -2u \quad . \tag{6.11}$$

(6.10) becomes:

$$u_t = u_{xxx} - 4u_x u - 2uu_x$$

$$= u_{xxx} - 6uu_x \quad , \tag{6.12}$$

which is the Korteweg-de Vries equation! This must be the simplest "derivation" of the Korteweg-de Vries equation in the Inverse Scattering context!

Let us then rewrite the basic equations as:

$$v_{xx} = uv \tag{6.13}$$

$$v_t = u_x v - 2uv_x \tag{6.14}$$

$$u_t = u_{xxx} - 6uu_x \quad . \tag{6.15}$$

Here is the problem: Given a solution (v,u) of (6.13)-(6.15), can we find a v' so that v' is a solution of (6.13)-(6.14), with a u' replacing u?

There are many ways of doing this; here is the way that reproduces the Estabrook-Wahlquist Bäcklund transformation [2].

Set:

$$v' = v^{-1} \tag{6.16}$$

$$v'_x = v^{-2} v_x$$

$$v'_{xx} = 2v^{-3}(v_x)^2 - v^{-2}v_{xx}$$

$$= 2v^{-3}(v_x)^2 - v^{-2}uv$$

$$= u'v'$$

$$= u'v^{-1} \quad ,$$

Hence,

$$u' = -u + 2\left(\frac{v_x}{v}\right)^2 \tag{6.17}$$

$$\left(\frac{v_x}{v}\right)_x = \frac{vv_x - (v_x)^2}{v^2}$$

$$= u - \left(\frac{v_x}{v}\right)^2 \tag{6.18}$$

Combine (6.17) and (6.18):

$$u' = -u + 2\left(u - \left(\frac{v_x}{v}\right)_x\right)$$

$$= u - \left(\frac{v_x}{v}\right)_x \quad ,$$

or

$$u - u' = \left(\frac{v_x}{v}\right)_x \quad . \tag{6.19}$$

So far, we have not touched Equation (6.14). Let us require that v satisfy it, with u' replacing u:

$$v'_t = -v^{-2}v_t$$

$$= -v^{-2}(u_x v - 2uv_x)$$

$$= u'_x v' - 2u'v'_x$$

$$= u'_x v^{-1} + \frac{2u'v_x}{v^2}$$

Hence,

$$u'_x + u_x = 2(u - u')\frac{v_x}{v} \tag{6.20}$$

Thus, u' must satisfy <u>both</u> (6.19) and (6.20). (In fact, notice that (6.20) follows from (6.17) and (6.19).)

It would now be convenient to eliminate v, i.e., to derive a set of differential equations linking u and u' <u>alone</u>. (This is what one usually means by a "Bäcklund transformation".) To do this, introduce new functions z, z' by the relation:

$$z_x = u \ ; \qquad z'_x = u' \quad . \tag{6.21}$$

(As we saw in Part B, this was also the transformation needed to put Korteweg-de Vries into <u>variational</u> form. See Whitham.) Choose the arbitrary constant involved here so that (6.19) integrates to give:

$$z - z' = \left(\frac{v_x}{v}\right) \quad . \tag{6.22}$$

Now, v_x/v satisfies a set of "Riccati"-like equations derived from the <u>linear</u> equations (6.13)-(6.14). These equations, when made explicit, are just the differential equations for z and z'! The formulas are worked out in the article by Chen [4], and I will not go into it here.

Bibliography

1. F. Estabrook and H. Wahlquist, "Prolongation Structures of Nonlinear Evolution Equations", J. Math. Phys. 16 (1975), pp. 1-7.

2. F. Estabrook and H. Wahlquist, "Bäcklund Transformations for Solutions of the Korteweg-de Vries Equation", Phys. Rev. Lett. 31 (1973), pp. 1386-1390.

3. H. Morris, "Bäcklund Transformation for Sine-Gordon", in Proceedings of the 1976 Ames (NASA) Conference on Geometric Nonlinear Waves, ed. by R. Hermann, to be published by Math Sci Press.

4. R. Muira, ed., "Bäcklund Transformations", Lecture Notes in Mathematics, Vol. 545, Springer-Verlag, 1974.

Chapter 2

SOME GEOMETRIC PROPERTIES OF THE SINE-GORDON BÄCKLUND TRANSFORMATION

1. INTRODUCTION

Usually, the Sine-Gordon equation and its Bäcklund transformation are viewed "locally". It actually has an interesting global structure as well (the physicists implicitly recognize this when they talk about "kinks"). Pohlmeyer [1] and Witten (to be published) have generalized some of its "inverse scattering properties" to the fields the physicists call the "sigma model". In this chapter I want to consider the Sine-Gordon Bäcklund transformation from a point of view which is very close to this work and also related to the Estabrook-Wahlquist "prolongation" ideas. I will begin with a brief study of the one and two-soliton solutions of the Sine-Gordon equation.

2. THE BÄCKLUND TRANSFORMATION AS AN SL(2,R) CONNECTION. RATIONALIZING COORDINATES

Let X be the space of variables

$$\phi, \phi_x, \phi_t \quad .$$

(Globally, X is $M^1(R^2, S_1)$, the first order mapping element (or "jet") space.) Let Y be the space of variables

$$(x, t, \phi, \phi_x, \phi_t, y) \quad .$$

Let $\pi: Y \to X$ be the map

$$\pi(x,t,\phi,\phi_x,\phi_t,y) = (x,t,\phi,\phi_x,\phi_t) \tag{2.1}$$

Then, π defines Y as a fiber space, with S_1 as fiber. Set:

$$\theta = dy - \left(\phi_x + 2\lambda \sin\left(\frac{\phi+y}{2}\right)\right) dx - \left(-\phi_t + \frac{2}{\lambda} \sin\left(\frac{y-\phi}{2}\right)\right) dt \tag{2.2}$$

θ defines an $SL(2,R)$-connection for the fiber space.

It will be convenient to change variables for spaces Y and X. How to do this is suggested by the formula for the "one-soliton" solution of (1.1), namely:

$$\tan \frac{u}{4} = C \exp\left(\frac{x-Ut}{\sqrt{1-U^2}}\right) , \tag{2.3}$$

with $\lambda = \sqrt{(1-U)/(1+U)}$. This suggests that we set:

$$v = \tan \frac{\phi}{4} \tag{2.4}$$

$$z = \tan \frac{y}{4} \tag{2.5}$$

Then,

$$d\phi = \frac{4}{1+v^2} dv \tag{2.6}$$

$$dy = \frac{4}{1+z^2} dz \tag{2.7}$$

$$\phi_x = \frac{4}{1 + v^2} v_x \tag{2.8}$$

$$\phi_t = \frac{4}{1 + v^2} v_t \tag{2.9}$$

Hence, new coordinates for X are

(x,t,v,v_x,v_t)

Coordinates for Y are

(x,t,v,v_x,v_t,z)

$$z = \frac{1}{i} \frac{e^{iy/4} - e^{-iy/4}}{e^{iy/4} + e^{-iy/4}}$$

$$e^{iy/4}(iz-1) = e^{-iy/4}(-1-iz) \quad ,$$

$$e^{iy/2} = \frac{1 + iz}{1 - iz} \tag{2.10}$$

$$e^{iu/2} = \frac{1 + iv}{1 - iv} \tag{2.11}$$

$$\sin\left(\frac{y+u}{2}\right) = \frac{1}{2i}\left(e^{i\left(\frac{y+u}{2}\right)} - e^{-i\left(\frac{y+u}{2}\right)}\right)$$

$$= \frac{1}{2i}\left(\left(\frac{1+iz}{1-iz}\right)\left(\frac{1+iv}{1-iv}\right) - \frac{(1-iz)}{(1+iz)}\frac{(1-iv)}{(1+iv)}\right)$$

$$= \frac{1}{2i}\left(\frac{(1+iz)^2(1+iv)^2 - (1-iz)^2(1-iv)^2}{(1+iz^2)(1+v^2)}\right)$$

$$= \frac{1}{2i}\left(\frac{(1+iz)^2(1+iv)^2-(1-iz)^2(1+iv)^2+(1-iz)^2(1+iv)^2-*1-iz)^2(1-iv)^2}{(1+z^2)(1+v^2)}\right)$$

$$= \frac{2z(1+iv)^2 + 2v(1-iz)^2}{(1+z^2)(1+v^2)}$$

$$= \frac{2(z+v-zv^2-vz^2)}{(1+z^2)(1+v^2)}$$

$$= \frac{2(z(1-v^2) + v(1-z^2)}{(1+z^2)(1+v^2)} \qquad (2.12)$$

Also,

$$\sin\frac{y-u}{2} = \frac{2(z(1-v^2) - v(1-z^2))}{(1+z^2)(1+v^2)} \qquad (2.13)$$

We can now make these substitutions into (2.2):

$$\theta = \frac{4}{(1+z^2)}\,dz - \left(\frac{4}{1+v^2}\,v_x + 4\lambda\left(\frac{z(1-v^2) + v(1-z^2)}{(1+z^2)(1+v^2)}\right)\right)dx$$

$$-\left(-\frac{4}{1+v^2}\,v_t + \frac{4}{\lambda}\left(\frac{z(1-v^2) - v(1-z^2)}{(1+z^2)(1+v^2)}\right)\right)dt$$

Hence, set:

$$\theta' \equiv \frac{1}{4}(1+z^2)(1+v^2)$$

$$= (1+v^2)dz - (v_x(1+z^2) + \lambda(z(1-v^2) + v(1-z^2))dx$$

$$- \left(-v_t(1+z^2) + \frac{1}{\lambda}(z(1-v^2) - v(1-z^2))dt\right) \qquad (2.14)$$

This is the definitive formula--it "rationalizes" the Sine-Gordon Bäcklund transformation in a very useful way, as we shall now see.

3. THE ONE-SOLITONS

Set

$$v = 0 \tag{3.1}$$

in (2.14). The result is

$$\theta'_0 = dz - \lambda z dx - \frac{1}{\lambda} z dt \quad ,$$

$$\frac{\theta'_0}{z} = d(\log z) - d\left(\lambda x + \frac{1}{\lambda} t\right) \tag{3.2}$$

Setting right hand side of (3.2) equal to zero, we have:

$$\log z = \lambda x + \frac{1}{\lambda} t + c$$

or

$$z = ce^{\lambda x + \frac{1}{\lambda} t} \tag{3.3}$$

which is, of course, the formula for one-solitons. Let us now substitute v = right hand side of (3.3) back into θ; the resulting solitons of $\theta = 0$ are called <u>two-solitons</u>.

4. THE TWO-SOLITONS

Set:

$$v = ce^{\left(\lambda x + \frac{1}{\lambda} t\right)} \tag{4.1}$$

$$v_x = \lambda v \; ; \qquad v_t = \frac{1}{\lambda} v \tag{4.2}$$

into (2.14). Call the resulting one-form θ_1:

$$\begin{aligned}\theta_1 &= (1+v^2)dz - (\lambda v(1+z^2) + \lambda(z(1-v^2) + v(1+z^2))dx \\ &\qquad - \left(-\frac{1}{\lambda} v(1+z^2) + \frac{1}{\lambda}(z(1-v^2) - v(1-z^2)\right)dt \\ &= (1+v^2)dz - \lambda(2v + z - zv^2)dx - \frac{1}{\lambda}(-2v + z - zv^2)dt\end{aligned} \tag{4.3}$$

Hence,

$$\theta_1 = (1+v^2)dz - (z-zv^2)d\left(\lambda x + \frac{t}{\lambda}\right) - 2vd\left(\lambda x - \frac{t}{\lambda}\right) \tag{4.4}$$

This is now in the form so that the equation $\theta_1 = 0$ is readily solvable by "quadratures". The explicit formula is in Whitham [ ], p. 609. Notice the physical meaning of the variables $x + \frac{t}{\lambda}$, $x - \frac{t}{\lambda}$. One represents "incoming waves", the other "outgoing waves". This interpretation suggests already how much more complicated will be the generalizations to wave equations in more independent variables. The expression on the right hand side of (4.4) will probably have to be replaced by some sort of "integral" over the "incoming" and "outgoing" directions.

5. THE SINE-GORDON BÄCKLUND TRANSFORMATION AS A "PSEUDO-POTENTIAL-PROLONGATION", I.E., CARTAN-EHRESMANN CONNECTION IN TWO VARIABLES

The global picture of what is going on is that we are dealing with connections for fiber spaces whose fibers are S_1, the unit circle in R^2. The "structure group" is SL(2,R) = SO(2,1). What we have done is to choose coordinates for the fiber in which the vector fields representing the action of the group take their "natural" form. We have seen that this is the coordinate system in which the equations for the one- and two-solitons take their most convenient form.

There is another approach which I shall now briefly develop, involving construction of connections with SO(2,1)-structure groups and R^2 as fiber. (SO(2,1) appears geometrically as the subgroup of the conformal group. SO(3,1), of R^2 which maps into itself the unit circle in R^2.) This provides a pseudopotential-prolongation structure for Sine-Gordon which is no doubt related to work by Pohlmeyer [1]. (I am indebted to E. Witten for this reference to Pohlmeyer's work and further discussions.)

Write two Pfaffian forms variables $(x,t,\phi,\phi',\phi_x,\phi_t,\phi'_x,\phi'_t)$:

$$\theta = d\phi - \left(\phi'_x + 2\lambda \sin\left(\frac{\phi+\phi'}{2}\right)\right) dx - \left(-\phi'_t + \frac{2}{\lambda} \sin\left(\frac{\phi-\phi'}{2}\right)\right) dt \tag{5.1}$$

$$\theta' = d\phi' - \phi'_x dx - \phi'_x dt \tag{5.2}$$

Introduce two sets of two real variables, i.e., coordinates for R^2:

$$y = (y^1, y^2) \ ; \quad z = (z^1, z^2) \quad .$$

Let

$$\cos\frac{\phi}{2} = y^1 \ ; \qquad \sin\frac{\phi}{2} = y^2$$

$$\cos \frac{\phi'}{2} = z^1 ; \quad \sin \frac{\phi'}{2} = z^2 .$$

Then,

$$\frac{\phi}{2} = \tan^{-1} \frac{y^2}{y^1}$$

$$\frac{d\phi}{2} = \frac{1}{1 + (y^2/y^1)^2} \left(\frac{y^1 dy^2 - y^2 dy^1}{(y^1)^2} \right)$$

$$= \left(\frac{y^1 dy^2 - y^2 dy^1}{(y^1)^2 + (y^2)^2} \right) \tag{5.3}$$

$$\sin\left(\frac{\phi+\phi'}{2}\right) = y^2 z^1 + y^1 z^2 \tag{5.4}$$

$$\sin\left(\frac{\phi-\phi'}{2}\right) = y^2 z^1 - y^1 z^2 \tag{5.5}$$

Hence,

$$= \frac{y^1 dy^2 - y^2 dy^1}{(y^1)^2 + (y^2)^2} - (\phi'_x + 2\lambda(y^2 z^1 + y^1 z^2))dx$$
$$- \left(-\phi'_t + \frac{2}{\lambda} (y^2 z^1 - y^1 z^2)\right) dt \tag{5.6}$$

Similarly,

$$\frac{\phi'_x}{2} = \frac{z^1 z^2_x - z^2 z^1_x}{(z^1)^2 + (z^2)^2} \tag{5.7}$$

$$\frac{\phi_t'}{2} = \frac{z^1 z_t^2 - z^2 z_t^1}{(z^1)^2 + (z^2)^2} \tag{5.8}$$

Define the following form in terms of the new variables:

$$\begin{aligned} \eta = \; & y^1 dy^2 - y^2 dy^1 - ((z^1 z_x^2 - z^2 z_x^1) + 2\lambda (y^2 z^1 + y^1 z^2) dx) \\ & - \left(-(z^1 z_t^2 - z^2 z_t^1) + \frac{2}{\lambda} (y^2 z^1 - y^1 z^2) \right) dt \end{aligned} \tag{5.9}$$

This form, restricted to the submanifolds

$$(y^1)^2 + (y^2)^2 = 1 \tag{5.10}$$

$$(z^1)^2 + (z^2)^2 = 1 \tag{5.11}$$

reproduces the old situation.

Now, with conditions (5.10)-(5.11), we have:

$$y^1 dy^1 + y^2 dy^2 = 0 \tag{5.12}$$

Substitute

$$dy^2 = \frac{-y^1}{y^2} dy^1 \tag{5.13}$$

into (5.9), using (5.10).

$$\begin{aligned} \eta = \; & \frac{-dy^1}{y^2} - ((z^1 z_x^2 - z^2 z_x^1) + 2\lambda (y^2 z^1 + y^1 z^2)) dx \\ & - \left(-(z^1 z_t^2 - z^2 z_t^1) + \frac{2}{\lambda} (y^2 z^1 - y^1 z^2) \right) dt \end{aligned}$$

Now set:

$$\theta^1 = dy^1 + (y^2((z^1 z_x^2 - z^2 z_x^1)) + 2\lambda(y^2 z^1 + y^1 z^2))dx + y^2\left(-(z^1 z_t^2 - z^2 z_t^1) + \frac{2}{\lambda}(y^2 z^1 - y^1 z^2)\right)dt \tag{5.14}$$

$$\theta^2 = -dy^2 + y^1((z^1 z_x^2 - z^2 z_x^1) + 2\lambda(y^2 z^1 + y^1 z^2))dx + y^1\left(-(z^1 z_t^2 - z^2 z_t^1) + \frac{2}{\lambda}(y^2 z^1 - y^1 z^2)\right)dt \tag{5.15}$$

This defines a "prolongation" in the sense of Estabrook and Wahlquist in two pseudopotential variables y^1 and y^2. In terms of the language of the theory of connections, θ^1, θ^2 define a connection for the fiber space whose fiber is R^2 and whose structure group is the group SO(2,1), the subgroup of the conformal group SO(3,1) of the flat Riemannian metric on R^2 which leaves invariant the spheres.

BIBLIOGRAPHY

1. K. Pohlmeyer, "Integrable Hamiltonian Systems and Interactions through Quadratic Constraints", Comm. Math. Phys.

2. G. Whitham, Linear and Non-Linear Waves, J. Wiley, New York, 1974.

Chapter 3

KINKS AND THE SINE-GORDON BÄCKLUND TRANSFORMATION

1. INTRODUCTION

Topology is one of the most active and vigorous branches of pure mathematics, which so far has had only a marginal impact on science and technology. The topological properties of solutions of non-linear differential equations must be more complicated and important to the study of the overall structure than for linear systems, hence--as science becomes more involved with nonlinear questions--emphasis on topology must increase. In particular, I believe that further topological information is crucial for progress in elementary particle, plasma-fusion physics; the physicist's lack of training and insight into geometric matters is deeply unfortunate here.

A great beginning was made twenty years ago by D. Finkelstein, Perring and Skyrme, with their theory of "kinks" of solutions of the Sine-Gordon equation and the other nonlinear field theoretic equations. Recently, elementary particle physicists have taken up these ideas again (finally!) and are more actively investigating certain topological objects ("monopoles", "instantons", etc.). In this chapter I want to contribute to this evolution by studying this traditional "kink" concept.

2. KINKS FOR SOLUTIONS OF THE SINE-GORDON EQUATION

Consider

$$\phi_{tt} - \phi_{xx} = \sin\phi \tag{2.1}$$

to be solved for a function

$$(x,t) \to \phi(x,t) \quad .$$

Regard ϕ as an <u>angle</u>. In order to do this, proceed as follows:

$$z = e^{i\phi} \quad ,$$

$$\phi = \frac{1}{i}\log z$$

$$i\phi_x = \frac{z_x}{z}$$

$$i\phi_{xx} = \frac{zz_{xx} - z_x^3}{z^2}$$

Thus, (2.1) becomes equivalent to the following system:

$$\frac{zz_{tt} - z_t^2 - zz_{xx} + z_x^2}{z^2} = \frac{1}{2}(z - z^{-1}) \tag{2.2}$$

$$|z| = 1 \quad . \tag{2.3}$$

Thus, if

$$(x,t) \to z(x,t)$$

is a solution of (2.2)-(2.3), the <u>kink</u>, $k(t)$, for fixed t, is the winding number of the curve

$$x \to z(x,t)$$

with respect to the origin is the complex w-plane as x goes from $-\infty$ to ∞, i.e.,

$$k(t) = \frac{1}{2\pi i} \int_{-\infty}^{\infty} \frac{z_x}{z}\, dx \tag{2.4}$$

<u>Remark</u>. This is the appropriate concept for "solitons". For periodic solutions, the limits of integration are taken over a period. The "solitons" are the limits of periodic solutions as the periods go to infinity.

3. THE KINK FROM THE POINT OF VIEW OF THE BÄCKLUND TRANSFORMATION

Instead of (2.1), work with the equivalent (after a change of variable, $(x,t) \to (\frac{1}{2}(x+t), \frac{1}{2}(x-t))$ equation

$$\phi_{xt} = \sin \phi \tag{3.1}$$

We know how to construct the Bäcklund transformation of (3.1). Write

$$\theta = d\psi - (\phi_x + 2\lambda \sin \frac{\phi+\psi}{2}\; dx - \left(-\phi_t + \frac{2}{\lambda} \sin\left(\frac{\psi-\phi}{2}\right)\right) dt \tag{3.2}$$

Now, set

$$z = e^{i\psi/2} \tag{3.3}$$

$$y = e^{i\phi/2} \tag{3.4}$$

$$\theta = \frac{2dz}{iz} - \left(\frac{2y_x}{iy} + \frac{2\lambda}{2i}(zy - z^{-1}y^{-1})\right) dx$$

$$- \left(- \frac{2y_t}{iy} + \frac{2}{\lambda 2i}(zy^{-1} - yz^{-1})\right) dt \tag{3.5}$$

Change independent variables as follows:

$$\begin{aligned} s_+ &= \lambda x + \frac{1}{\lambda} t \\ s_- &= \lambda x - \frac{1}{\lambda} t \end{aligned} \tag{3.6}$$

$$\begin{aligned} 2\lambda x &= s_+ + s_- \\ \frac{2}{\lambda} t &= s_+ - s_- \end{aligned} \tag{3.7}$$

The one-soliton solutions are found by setting

$$y = 1$$

$$\theta_1 = \frac{2dz}{iz} - \left(\frac{\lambda}{i}(z-z^{-1})\right)dx - \left(\frac{1}{\lambda i}(z-z^{-1})\right)dt$$

$$= \frac{2dz}{iz} - \frac{1}{i}(z-z^{-1})ds_+ \tag{3.8}$$

The functions $z(s_+,s_-)$ which annihilate θ, i.e.,

$$\frac{dz}{ds^+} = \frac{1}{2}(z^2-1) \tag{3.9}$$

are the <u>one-solitons</u>.

The right hand side of (3.9) defines a vector field on the unit circle in $\mathbb{C}$. (It belongs to the Lie algebra of $U(1,1)$ acting on the circle.) Its zero points are at

$$z = \pm 1 \tag{3.10}$$

These correspond to the solution $\psi = 0$ or π of Sine-Gordon, i.e., kink number zero.

At $z = \pm 1$, the linearization of the vector field is

$$\pm z \quad .$$

Thus, $z = \pm 1$ is an <u>attractor</u> for the vector field, $z = -1$ is a <u>repulsor</u>. Thus, a solution $z(s_+)$ of (3.9) which does not satisfy (3.10) will behave as follows as $s_+ \to \infty$

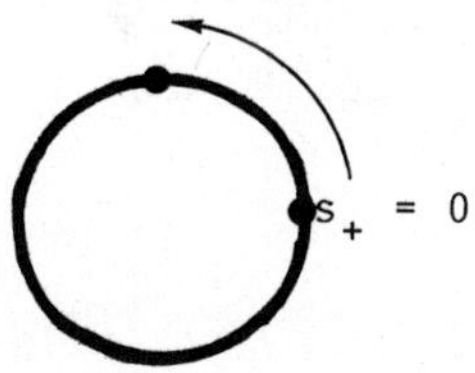

Similarly, as $s_+ \to -\infty$

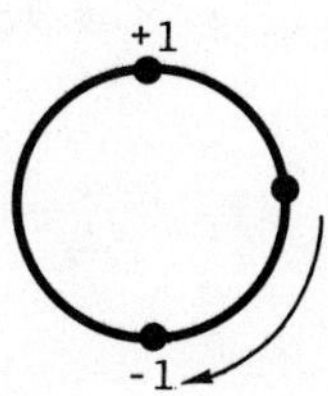

Putting the two together, we get a curve $s_+ \to z(s_+)$ which goes monotonically from -1 at $s_+ = -\infty$ to -1 at $s_+ = +\infty$. Since $z = e^{i\psi/2}$, this corresponds to changing by 2π, i.e., the <u>kink number of these one-soliton solutions is one</u>. (Notice the geometric formulation, in terms of vector fields on the compact manifold S^1.)

Now, let us look at the two-solitons. Let $y(x,t)$ be a one-soliton solution, i.e.,

$$
\begin{aligned}
y_x &= \frac{1}{2}(y^2 - 1)\lambda \\
y_t &- \frac{1}{2\lambda}(y^2 - 1) \quad .
\end{aligned}
\tag{3.11}
$$

Insert this into θ, obtaining the following form:

$$
\begin{aligned}
\frac{i\theta_2}{2} &= \frac{dz}{z} - \left(\frac{(y^2-1)}{2y}\lambda + \frac{\lambda}{2}(zy - z^{-1}y^{-1})\right) dx \\
&\quad - \left(-\frac{(y^2-1)}{2\lambda y} + \frac{1}{2\lambda}(zy^{-1} - yz^{-1})\right) dt \\
&= \frac{dz}{z} - \frac{1}{2}(zy - z^{-1}y^{-1})ds_+ - \frac{y^2-y}{2y}ds^-
\end{aligned}
\tag{3.12}
$$

Hold s_- fixed. The function $z(s_-)$ resulting from setting $\theta_2 = 0$ now satisfies the following ordinary differential equation:

$$
\frac{dz}{ds} = \frac{1}{2}(z^2y - y^{-1}) \quad . \tag{3.13}
$$

Now z is on an orbit of an s-dependent vector field with the circle in $\mathbb{C}$. y satisfies a differential equation of its own, namely:

$$\frac{dy}{ds} = \frac{1}{2}(y^2-1) \quad . \tag{3.14}$$

(3.11) and (3.12) tell us that $s \to (y(s),z(s))$ is a curve in $\mathbb{C}^2$ which is the orbit curve of a vector field, namely,

$$A = \frac{1}{2}(y^2-1)\frac{\partial}{\partial y} + \frac{1}{2}(z^2 y - y^{-1})\frac{\partial}{\partial z} \tag{3.15}$$

We are now interested in how many times the orbit curves pass through

$$z = 1 \quad .$$

This, counted with appropriate multiplicity, will be the kink.

It is also interesting to observe that (3.15) defines A as an element of the <u>finite dimensional</u> Lie algebra of vector fields on $\mathbb{C}^2$. Thus, we have the possibility of computing the two-solitons and the kinks by purely Lie group-theoretic methods. This might be a significant tie-in between the theory of solitons and Lie group theory!

To investigate the asymptotic behavior of the orbits of A, let us first find its zero points.

$$\begin{aligned} y &= \pm 1 \\ z &= \pm 1 \quad . \end{aligned} \tag{3.16}$$

Let us linearize (3.11)-(3.12) in the neighborhood of the zero points (3.14).

$$\frac{d(\delta y)}{ds} = y\delta y = \pm\,\delta y$$

$$\frac{d(\delta z)}{ds} = z\delta zy + \frac{1}{2}\,z^{2}\delta y + y^{-2}\delta y \tag{3.17}$$

$$= (\pm\pm)\,\delta z \pm \frac{1}{2}\,\delta y - \delta y$$

In order to determine the stability properties of the zero points (3.17), we must examine the eigenvalues of the matrix of the linear system (3.17). This matrix is

$$\begin{pmatrix} 1 & 0 \\ \pm\frac{1}{2} - 1 & (\pm 1)(\pm 1) \end{pmatrix}$$

We are looking for the attractor zero point, i.e., the one of those listed in (4.16) whose eigenvalues are <u>negative</u>. We see that it must be

$$y = -1, \qquad z = +1 \tag{3.18}$$

Similarly, the repulsive critical point is that for which

$$y = +1, \qquad z = +1 \quad . \tag{3.19}$$

Let us suppose that the following behavior is typical of a solution $s \to (y(s), z(s))$ of the system (3.11)-(3.12):

$$\lim_{s\to+\infty} (y(s),z(s)) = (-1,+1)$$

$$\lim_{s\to-\infty} (y(s),z(s)) = (+1,-1)$$

We see that, for ψ such that

$$z = e^{i\psi/2}$$

$\psi(s)/2$ goes from 0 to 2π, i.e., $\psi(s)$ goes from 0 to 4π, i.e., the kink number of the two-soliton solution should be two.

Of course, this is not a completely rigorous proof--but it is an interesting intuitive way of tying up the kink number with qualitative properties of the Bäcklund transformation. In fact, we see that the kink number is probably closely tied up with certain singularity properties of the Pfaffian equation defining the Bäcklund transformation.

Chapter 4

MULTIPLE SOLITONS, MULTIPROLONGATIONS, AND THE BIANCHI PERMUTATION THEOREM

1. INTRODUCTION

The property of the Sine-Gordon Bäcklund transformation that is called the Bianchi permutation theorem is one of the most remarkable. As emphasized particularly by G. Lamb [1], it plays a key role in various physical applications.

I believe (along with Hugo Wahlquist and Frank Estabrook who suggested this problem to me) that the phenomenon has a Lie group or differential geometric meaning. This is certainly plausible, in view of the close relation between the Bäcklund transformations and connections with Lie groups as structure groups. I also believe that it has something to do with the "addition theorems" for elliptic functions. At any rate, in this chapter I intend to vigorously (but tentatively) pursue the geometric background to this material.

2. SINGLE PSEUDOPOTENTIAL VARIABLE-QUADRATIC PROLONGATIONS

Adopt (x,t) as independent variables, u as a single dependent variable. Let X be the space of variables

$$(x,t,u,u_x,u_t,\cdots) \tag{2.1}$$

This X is an appropriate "jet" (or "mapping element") space of mappings $R^2 \to R$. Also, informally, X is parameterized by x, t, u and various derivatives.

Let

$$\pi: Y \to X$$

be a fiber space (i.e., a submersion) with a one-dimensional fiber, whose fiber coordinate is denoted by y. (We usually work locally although the results can be interpreted globally.) Make no notational distinction between the functions $(x,t,u,u_x,\ldots)$ on X and their pull-back under π to be functions on Y.

Suppose that θ is a one-form on Y of the following type.

$$\theta = dy - \theta_0 - \theta_1 y - \theta_2 y^2 \tag{2.2}$$

where $\theta_0, \theta_1, \theta_2$ are one-forms which are pull-backs under π of one-forms on X. As explained in Parts A and B, (2.2) should be interpreted as an $SL(2,R)$ connection, with θ_0, θ_1, θ_2 defining a Lie-algebra-valued one-form on X. In non-linear wave theory, this is called a <u>single pseudopotential, quadratic prolongation</u>. The prolongation for Korteweg-de Vries, Sine-Gordon, non-linear Schrödinger,... seems to be of this form. (However, the Boussinesq equation seems to require a more complicated one.)

Suppose that the ω_i, $i = 1,2,3$, can be written as:

$$\omega_i = a_i(u,u_x,u_t)dx + b_i(u,u_x,\ldots)dt \tag{2.3}$$

Consider a pair $(u(x,t), y(x,t))$ of functions which annihilate θ, i.e., which are such that:

$$
\begin{aligned}
y_x &= a_0(u(x,t),\ldots) + a_1(\ )y + a_2(\ )y^2 \\
y_t &= b_0(\qquad\qquad) + b_1(\ )y + b_2(\ )y^2
\end{aligned}
\tag{2.4}
$$

3. ITERATED PROLONGATIONS

Consider (2.4). Begin with a $u(x,t)$ which makes the one-form θ completely integrable (in the Frobenius sense, i.e., annihilates the curvature forms). (2.4), an equation for $y(x,t)$, is then completely integrable.

Suppose now that we feed y into another single pseudopotential quadratic prolongation, with "z" the pseudopotential variable. This means that we construct another space Z with coordinates

$$(x,t,y,y_x,y_t,\ldots,z) \quad ,$$

and a one-form:

$$\omega = dz + \omega_0 + \omega_1 z + \omega_2 z^2 \quad , \tag{3.1}$$

where $\omega_0,\omega_1,\omega_2$ are one-forms in the variables $(x,t,y,y_x,\ldots)$ of the form:

$$\omega_i = \alpha_i(y,y_x,\ldots)dx + \beta_i(y,y_x,\ldots)dt \tag{3.2}$$

Now, (3.2) gives us the values of $y_x,y_t,\ldots$ in terms of

$u, u_x, \ldots$. Let us substitute them into (3.2), obtaining forms that we denote by

$$
\begin{aligned}
\tilde{\omega}_i &= \alpha_i(y(u,u_x,\ldots)dx + \beta_i(\ )dt \\
&= \tilde{\alpha}_i(u,u_x,\ldots)dx + \tilde{\beta}_i(u,u_x,\ldots)dt
\end{aligned} \tag{3.3}
$$

Thus, we see that the $z(x,t)$ which annihilate (3.1), together with the $y(x,t)$, satisfies:

$$dy + \theta_0 + \theta_i y + \theta_2 y^2 = 0 \tag{3.4}$$

$$dz + \tilde{\omega}_0 + \tilde{\omega}_1 y + \tilde{\omega}_2 y^2 = 0 \tag{3.5}$$

Consider a fiber space on X with a two-dimensional fiber that we denote as (y,z). We see that (3.4) defines connections for this fiber space with $SL(2,R) \times SL(2,R)$ as structure group. In the next chapter we shall work out this construction--including the global situation--in much more detail. At the moment, it is appropriate to study the main example--the Sine-Gordon prolongations.

4. THE SINE-GORDON PROLONGATIONS

Start off with the standard Sine-Gordon Bäcklund transformation:

$$
\begin{aligned}
y_x &= u_x + 2\lambda_1 \sin\left(\frac{y+u}{2}\right) \\
y_t &= -u_t + \frac{2}{\lambda_1} \sin\left(\frac{y-u}{2}\right)
\end{aligned} \tag{4.1}
$$

(We have seen that it is easy to change variables to make this a quadratic prolongation. It will be more convenient to work with the trigonometric functions.)

Thus, let us write the transformation from y to z as a Pfaffian equation:

$$dz = \left(y_x + 2\lambda_2 \sin\left(\frac{z+y}{2}\right)\right) dx + \left(-y_t + \frac{2}{\lambda_2} \sin\left(\frac{z-y}{2}\right)\right) dt$$

$$= \text{, after substitution of (4.1)}$$

$$\left(u_x + 2\lambda_1 \sin\left(\frac{y+u}{2}\right) + 2\lambda_2 \sin\left(\frac{z+y}{2}\right)\right) dx$$

$$+ \left(u_t - \frac{2}{\lambda_1} \sin\left(\frac{y-u}{2}\right) + \frac{2}{\lambda_2} \sin\left(\frac{z-y}{2}\right)\right) dt$$

$$= du + \left(2\lambda_1 \sin\left(\frac{y+u}{2}\right) + 2\lambda_2 \sin\left(\frac{z+y}{2}\right)\right) dx + \left(\frac{2}{\lambda_2} \sin\left(\frac{z-y}{2}\right) + \frac{2}{\lambda_1} \sin\left(\frac{u-y}{2}\right)\right) dt \qquad (4.2)$$

Also,

$$dy = \left(u_x + 2\lambda_1 \sin\left(\frac{y+u}{2}\right)\right)dx + \left(-u_t + \frac{2}{\lambda_1} \sin\left(\frac{y-u}{2}\right)\right)dt \qquad (4.3)$$

Interpret the space of variables (x,t,u,u_x,u_t,y,z) as a fiber space over the space of variables (x,t,u,u_x,u_t), with a two-dimensional fiber. Formulas (4.2) and (4.3) then define Cartan-Ehresmann connections for this fiber space with SL(3,R) as structure group (acting via projective transformations on R^2.)

5. THE BIANCHI THEOREM OF PERMUTABILITY

In the classical literature, this circle of ideas is most confusing. There is a recent very clear presentation in the paper "A Restricted Bäcklund Transformation", J. Math. Phys. 14 (1973), 1817-1828, by McLaughlin and Scott which I will follow.

Suppose that we start off with a function $u(x,t)$ which satisfies Sine-Gordon. Let λ_1, λ_2 be two distinct values of the parameter that appears in the Bäcklund transformation. Let y, y' be Bäcklund transformations of u with values of the parameters λ_1, λ_2.

Theorem 5.1. There is a function $F(\ , \ , \ , \ )$ of four variables such that whatever $y, y', \lambda_1, \lambda_2$,

$$z(x,t) = F(y(x,t), y'(x,t), \lambda_1, \lambda_2) + u(x,t) \quad (5.11)$$

is a Bäcklund transformation of y with λ_2 as value of its parameter <u>and</u> a Bäcklund transformation of y' with λ_1 as parameter value. This can be symbolized by the following "commutative Lamb diagram".

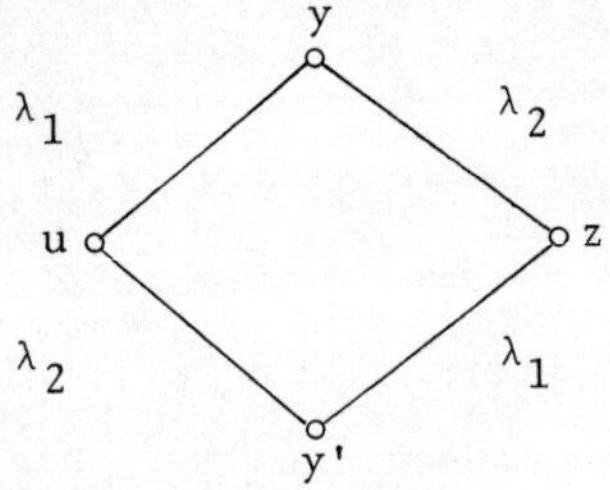

Further, F is "symmetric" in the sense that

$$F(y,y',\lambda_1,\lambda_2) = F(y',y,\lambda_2,\lambda_1) \qquad (5.2)$$

<u>Proof</u>. This is proved in detail in the paper cited above where the following formula is given for F:

$$F(y,y',\lambda_1,\lambda_2) = \tan^{-1}\left(\frac{\lambda_1+\lambda_2}{\lambda_1-\lambda_2} \tan\left(\frac{y-y'}{4}\right)\right) \qquad (5.3)$$

I will refer there for the proof.

<u>Remark</u>. A casual glance at the classical literature might lead one to believe that this result is related to the "commutativity" property of groups. However, the Bäcklund transformation is <u>not</u> a group, but something more general. In a later section I will describe some abstract ideas which enable one to think of the Bianchi theorem as a <u>generalization</u> of the commutativity property.

The theorem is also much more widely useful for the practical problem of solving the Sine-Gordon equation than appears at first sight. For, if one <u>special</u> solution to a Bäcklund relation is found, all the others can be found by "quadratures". (This is a general property of Bäcklund transformation which are defined by means of <u>single</u>, <u>quadratic prolongations</u> in the sense of Estabrook and Wahlquist. For the solutions of such equations are determined by <u>Riccati</u> equations and they (classically) have the property that if one particular solution is known, all others can be found by "quadrature.")

6. THE BUNDLE INTERPRETATION OF THE BIANCHI THEOREM

We can now give a very elegant geometric interpretation of the Bianchi relation. This formulation will suggest many generalizations, global interpretations, etc.

Let X be the space of variables

$$(x,t,u,u_x,u_t) \quad .$$

Construct two fiber spaces Y,Z over X with two-dimensional fibers. The fiber variables for Y are labelled

$$y_1, y_2$$

The fiber variables for Z are labelled

$$(y,z) \quad .$$

Let (λ_1,λ_2) be additional real parameters. Define Ehresmann connections for each of these bundles by giving the following connection forms:

$$\begin{aligned} \theta_1 = dy_1 &- \left(u_x + 2\lambda_1 \sin\left(\frac{y_1+u}{2}\right)\right) dx \\ &- \left(-u_t + 2\lambda_1^{-1} \sin\left(\frac{y_1-u}{2}\right)\right) dt \end{aligned} \qquad (6.1)$$

$$\theta_2 = dy_2 - \left(u_x + 2\lambda_2 \sin\left(\frac{y_2+u}{2}\right)\right) dx$$
$$- \left(- u_t + 2\lambda_2^{-1} \sin\left(\frac{y_2-u}{2}\right)\right) dt \qquad (6.2)$$

This defines an

SL(2,R) × SL(2,R)

connection for the fiber space $Y \to X$.

Now, define a connection for Z by the following one-form:

$$\omega_1 = dy - \left(u_x + 2\lambda_1 \sin\left(\frac{y+u}{2}\right)\right) dx$$
$$- \left(- u_t + 2\lambda_1^{-1} \sin\left(\frac{y-u}{2}\right)\right) dt \qquad (6.3)$$

$$\omega_2 = du + 2\left(\lambda_1 \sin\left(\frac{y+u}{2}\right) + \lambda_2 \sin\left(\frac{z+u}{2}\right)\right) dx$$
$$+2\left(\lambda_2^{-1} \sin\left(\frac{z-y}{2}\right) + \lambda_1^{-1} \sin\left(\frac{u-y}{2}\right)\right) dt \qquad (6.4)$$

As we have already remarked, this is (often an appropriate "rationalizing" change of variables) an SL(3,R)-connection, where SL(3,R) acts on R^2 via <u>projective</u> transformation.

Now, define a fiber preserving map

$\phi: Y \to Z$

be the following formula:

$$(x,t,u,u_x,u_t,y_1,y_2)$$

$$= \left(x,t,u,u_x,u_t,y = y_1,\ z = \tan^{-1}\left(\frac{\lambda_1+\lambda_2}{\lambda_1-\lambda_2}\right) \tan\left(\frac{y_1-y_2}{4}\right) + u\right) \tag{6.5}$$

The Bianchi theorem now can be described precisely as follows.

> Let S be a two-dimensional submanifold of Y on which $dx \wedge dt \neq 0$ and such that the connection forms θ_1, θ_2 and the contact form $du - u_x dx - u_t dt$ are zero when restricted to S. (S is then determined by a solution of Sine-Gordon plus a pair $(y_1(x,t),\ y_2(x,t))$ of Bäcklund transforms of u for parameters λ_1, λ_2.) Then, the connection forms ω_1, ω_2 on the contact forms $du - u_x dx - u_t dt$ are zero when restricted to $\phi(S)$. (6.6

I believe that this result can be sharpened to prove the following formula (which would imple (6.6):

> $\phi^*(\omega_1, \omega_2)$ are linear combinations of θ_1, θ_2 and $du - u_x dx - u_t dt$. (6.7

(6.7) would say that:

ϕ <u>is a connection isomorphism</u>.

However, I have not had the calculational courage to check whether (6.7) is actually verified. I leave it to the ambitious reader to enquire into this point!

7. AN ABSTRACT LOOK AT THE "THEOREM OF PERMUTATIVITY" FROM THE POINT OF VIEW OF "CORRESPONDENCES"

Bäcklund transformations are not maps, hence it is not a priori meaningful to consider their "compositions". However, as "correspondences", one can attempt to look for generalizations of concepts that make sense for maps. In particular, for "groups" one has the transformation of a "one-parameter group of transformations" as a homomorphism map of the additive group of the real numbers into the group (under composition) of transformations on a space.

Let $\mathcal{S}, \mathcal{S}'$ be sets. Consider

$$\pi: \mathcal{S} \times \mathcal{S}' \to \mathcal{S}$$

$$\pi': \mathcal{S} \times \mathcal{S}' \to \mathcal{S}'$$

as the Cartesian projection. A subset

$$\mathcal{C} \subset \mathcal{S} \times \mathcal{S}'$$

is called a correspondence form $\mathcal{S} \to \mathcal{S}'$ if

$$\pi\mathcal{C} = \mathcal{S} \tag{7.1}$$

Given $s \in \mathcal{S}$, the set of $s' \in \mathcal{S}'$ such that

$$(s,s') \in \mathcal{C}$$

is called the <u>image</u> of s under $\mathcal{C}$, and denoted by

$$\mathcal{C}(s) \quad .$$

Note that

$$\mathcal{C}(s) = \pi'(\pi^{-1}(s) \cap \mathcal{C}) \quad . \tag{7.2}$$

Thus, a <u>correspondence</u> may be thought of as a <u>map</u> in the ordinary sense

$$\mathcal{S} \to (\text{subsets of } \mathcal{S}') \quad .$$

Typically, one imposes other "functional" conditions on $\mathcal{C}$ to make it interesting. For example, in algebraic geometry $\mathcal{S}$ and $\mathcal{S}'$ may be "algebraic sets". One would then naturally require that $\mathcal{C}$ be an "algebraic subset" of $\mathcal{S} \times \mathcal{S}'$. In the 19th century work, there was an analogy in the minds of many of the workers in the field of differential equations between <u>differential</u> and <u>algebraic equations</u>. Thus, in terms of this analogy, a "Bäcklund transformation" is the analogue of an "algebraic correspondence" in algebraic geometry.

Let us continue with the abstract theory. Let $\mathcal{S}$, $\mathcal{S}'$, $\mathcal{S}''$ be spaces and let

$$\mathcal{C} \subset \mathcal{S} \times \mathcal{S}' \ ; \qquad \mathcal{C}' \subset \mathcal{S}' \times \mathcal{S}''$$

be correspondences $\mathscr{S} \to \mathscr{S}'$, $\mathscr{S}' \to \mathscr{S}''$. We construct the <u>composite</u> correspondence

$$\mathscr{C}'' \equiv \mathscr{C}'\mathscr{C} \tag{7.3}$$

as the set of pairs

$$(s,s'') \in \mathscr{S} \times \mathscr{S}''$$

such that there exists an $s' \in \mathscr{S}'$ such that

$$(s,s') \in \mathscr{S} \; ; \qquad (s',s'') \in \mathscr{S}' \quad .$$

In other words, $\mathscr{C}'\mathscr{C}$ is the obvious generalization of the notion of composition of maps. At the level of "sets", this is just a composition of maps in the usual sense.

$$(\mathscr{C}'\mathscr{C})(s) = \mathscr{C}'(\mathscr{C}(s)) \tag{7.4}$$

Suppose now that

$$\mathscr{S} = \mathscr{S}' = \mathscr{S}'' \quad .$$

Let us say that correspondences

$$\mathscr{C}: \mathscr{S} \to \mathscr{S} \quad ; \qquad \mathscr{C}': \mathscr{S} \to \mathscr{S}$$

<u>commute</u> if, for each $s \in \mathscr{S}$, the subsets

$$\mathscr{C}'(\mathscr{C}(s)) \quad \text{and} \quad \mathscr{C}(\mathscr{C}'(s))$$

<u>have an element in common</u>.

We can now state the Bianchi theorem in this abstract framework. Let $\mathscr{S}$ = space of solutions of Sine-Gordon equation. For each λ, let

$$\mathcal{B}_\lambda = \{(s,s') \in \mathcal{S} \times \mathcal{S} : \text{ s and s' are related via a Bäcklund transformation with parameter value } \lambda\} \quad .$$

The Bianchi theorem now says (in part) that

$$\lambda \to \mathcal{B}_\lambda$$

forms a <u>one-parameter commutative family of Bäcklund transformations</u>.

I suspect that these abstract ideas may play a significant role in physics; possibly even in the foundations of quantum mechanics.

BIBLIOGRAPHY

1. G. Lamb, "Bäcklund Transformations at the Turn of the Century", in <u>Bäcklund Transformations</u>, R. Muira, ed., Lecture Notes in Mathematics, Vol. 515, Springer-Verlag, 1976.

2. D. McLaughlin and A. Scott, <u>J. Math. Phys</u>. <u>14</u> (1973), 1817-1828.

Chapter 5

SOME CONCEPTS FROM THE GENERAL THEORY OF FIBER SPACES AND CARTAN-EHRESMANN CONNECTIONS

1. INTRODUCTION

We are slowly coming to the realization in this series that the theory of nonlinear waves-solitons involves fundamental new material in the theory of fiber spaces and connections. Certainly, I have found the formulation of the ideas in this framework to be uncannily relevant!

At some point, I will want to investigate the exact _global_ nature of the bundles constructed in terms of Estabrook-Wahlquist prolongations. (For example, I think the explication of the _kink_ notion lies in this direction.) As preparation, I will now gather together certain general material that will be useful here. Of course, this material is also useful in broader areas of mathematical physics, e.g., in the study of "gauge fields" (which are just "connections").

2. FIBER SPACES AS SUBMERSIONS

All spaces will be smooth manifolds, all maps smooth, etc., unless mentioned otherwise. Usually, the degree of smoothness will be C^∞. If X is a manifold, $T(X)$ denotes its tangent bundle. For $x \in X$, X_x denotes the tangent space to X and x.

A map $\phi: Y \to X$ between manifolds will be called a <u>submersion</u> if the induced map on tangent vectors

$$\phi_*: T(Y) \to T(X)$$

is <u>onto</u>. The implicit function theorem then implies the plentiful local existence of open subsets

$$O \subset Y$$

$$O' \subset X$$

such that:

$$\phi(O) = O'$$

O is diffeomorphic to the product $O' \times Z$ of O' and a "fiber" Z, such that becomes the Cartesian projection

$$O' \times Z \to O' \quad .$$

In the language of coordinates, this means that Y has local coordinates of the form $(\phi^*(x), z)$, where (x) are coordinates for X, and z are a set of functionally independent functions. (They will be called <u>fiber coordinates</u> for ϕ.)

There are many possible definitions of "fiber space" and "fiber bundle" in the literature. The concept is rather flexible, and it is not desirable to be rigid, but to adapt the definition to the desired purpose. For the purpose of this chapter it is useful to <u>define a fiber space as a triple</u> (Y,X,ϕ), <u>where</u> $\phi: Y \to X$ <u>is a submersion map</u>.

If (Y,X,ϕ) is such a fiber space, and if O is an open subset of X, $\Gamma(O)$ denotes the space of maps $\gamma: O \to Y$ that are cross-section maps for ϕ, i.e., such that

$$\phi\gamma = \text{identity} \quad .$$

The Implicit Function Theorem guarantees a plentiful supply of such cross-sections, if O is a sufficiently small open subset of X. Here is the relevant picture.

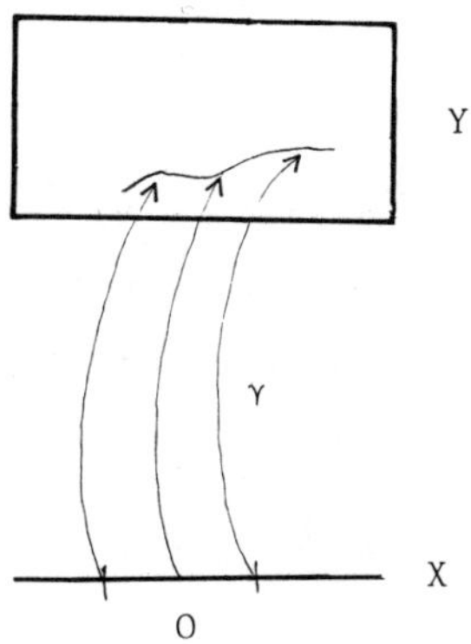

Our plan is to study the imposition of various <u>geometric</u> structures on the manifolds and fiber spaces by means of certain <u>algebraic</u> structures in the $\Gamma(O)$. (For example, a <u>vector bundle</u> structure means that $\Gamma(O)$ is a <u>vector space</u>.)

3. BUNDLES OF LIE GROUPS

Let

$$\pi\colon Y \to X$$

be a fiber space. For each $x \in X$, let

$$Y(x) = \pi^{-1}(x)$$

be the fiber above the point x. It is a (regularly embedded) submanifold of Y. If each fiber $Y(x)$ is _given a Lie group structure_, _and this Lie group multiplication varies smoothly with_ x, then we say that we are given a _bundle of Lie groups_.

The "smoothness in x" property can be described alternatively as follows:

> Given $\gamma_1, \gamma_2 \in \Gamma(0)$, for $0 \subset X$, i.e., two cross-sections over an open subset 0 of the base space X, we can use the group structure on the fibers to "multiply" γ_1 and γ_2:
>
> $$(\gamma_1\gamma_2)(x) = \gamma_1(x)\gamma_2(x) \qquad (3.1)$$
>
> for $x \in X$.
>
> "Smoothness" requires that $\gamma_1\gamma_2$ so defined be smooth also, i.e., lie in $\Gamma(0)$. This makes $\Gamma(0)$ into a group. (Physicists call this _the gauge group_.)

The reader should keep in mind that there are certain differences between this concept and that of <u>principal bundle</u> in the theory of fiber bundles. This is defined as a fiber space $\pi: Y \to X$ and an action of a Lie group G on Y so that:

a) The orbits of G are exactly the fibers of π

b) G acts in a simple way on Y, i.e., if $gy = y$ for some $y \in Y$, $g \in G$, then g = identity. In other words, the orbits of G are identified with the fibers of π.

However, this identification of G with the orbits of π is <u>not unique</u>. Hence, it is not possible to define the fibers uniquely as a Lie group--this is the difference! Of course, if a cross-section, say $\phi: X \to Y$ is <u>given</u>, this identification is possible, and the Y becomes a "bundle of groups". This requires that Y be a product of X and G. Locally then, "bundles of groups" and "principal bundles", but there may be global differences.

4. BUNDLES OF LIE GROUPS ACTING ON FIBER SPACES. BUNDLES OF TRANSFORMATION GROUPS

Let

$$\pi: Y \to X \; ; \qquad \pi': Z \to X$$

be two fiber spaces, with the same base space X. Suppose that (Y,X,π) is a bundle of Lie groups.

For each $x \in X$, let $Y(x) = \pi^{-1}(x)$, $Z(x)$ be the fibers of each above x. $Y(x)$ is given to us as a Lie group. We shall suppose that a transformation group action is given if a $Y(x)$ on $Z(x)$, i.e., a map

$$Y(x) \times Z(x) \rightarrow Z(x) \quad , \tag{4.1}$$

denoted by $(y,z) \rightarrow yz$ is given satisfying the usual transformation group conditions. Of course, this transformation group action "varies with x". In order to define its <u>smoothness</u> as x varies, proceed as follows. Let O be an arbitrary open subset of X. $\Gamma(O)$, $\Gamma'(O)$ denote the spaces of smooth cross-section maps

$$O \rightarrow Y \; ; \qquad O \rightarrow Z \quad .$$

Using (4.1) we define the action of $\gamma \in \Gamma(O)$ on $\gamma' \in \Gamma'(O)$:

$$\gamma\gamma'(x) = \gamma(x)(\gamma'(x)) \quad . \tag{4.2}$$

We now <u>require</u> that (4.2) define $\gamma\gamma'$ as a <u>smooth</u> element of $\Gamma'(O)$. This fixes the intuitive idea that the "transformation group action of the group $Y(x)$ on $Z(x)$ varies smoothly with x".

Here is one non-trivial example of such a set up. Suppose G is a group, V a vector space, X a space, and that for each $x \in X$, we have a linear representation

$$\rho_x \colon G \rightarrow L(V)$$

of G by linear maps in V. Set:

$$Y = X \times G \quad ; \quad \pi: Y \to X$$

the Cartesian projection.

$$(x,g_1)(x,g_2) = (x,g_1g_2) \quad .$$

Thus, Y is the _product_ bundle. Similarly, define

$$Z = X \times V$$

as the _product_ bundle. However, set

$$(x,g)(x,v) = (x,\rho_x(g)(v))$$

$$\text{for } g \in G, \quad v \in V \quad .$$

This defines an action of the bundle of groups on the bundle of vector spaces $X \times V$ that is truly "twisted" in x. It will inherit whatever "smoothness" properties the map $(x,g) \to \rho_x(g)$ is endowed with.

The moral of this "example" is that this is a vast topic--including much of what is known as "analytic continuation" or "deformations" or "contractions" of group representation, group actions, etc. See my series of papers, "Analytic Continuation of Group Representations" for an introduction to this circle of ideas. This material is also extremely relevant to elementary particle physics-quantum field theory, via such topics as "current algebras". There is extensive material developed in my books _Lie Algebras and Quantum Mechanics_, _Vector Bundles in Mathematical Physics_, _Geometry, Physics and Systems_ and Interdisciplinary Mathematics, Volume VI about this topic.

5. CARTAN-EHRESMANN CONNECTIONS

Material with this title was developed in "Interdisciplinary Mathematics, Volume X". Here I want to briefly recall the definitions, then describe its relation to the theory of "bundles of groups acting on fiber spaces". Of course, my reason for going over this again here is that I want to interpret "prolongation" in the sense of Estabrook and Wahlquist as <u>globally defined</u> Cartan-Ehresmann connections, then associate the "kink" invariants with certain geometric objects associated with these connections.

Let

$$\pi: Z \to X$$

be a fiber space. Let $\mathcal{V}$ be the smallest $F(Z)$-module of $F^1(Z)$ containing the forms

$$\pi^*(F^1(X)) \quad .$$

The one-forms in $\mathcal{V}$ are said to be the <u>verical</u> one-forms with respect to the map .

<u>Definition</u>. An <u>Ehresmann connection</u> is defined by an $F(X)$-submodule of $F^1(Z)$ --denoted typically by E --such that

> The $F(Z)$-module $F^1(Z)$ is a <u>direct sum</u> (in the module sense) of $\mathcal{V}$ and E.

The one-forms in E are called the <u>horizontal forms</u>.

Here is what this means in terms of local coordinates. Let

$$(x^i) \quad , \qquad 1 \leq i,j \leq n$$

be a coordinate system for X. Make no notational distinction between them and their pull-backs to Z under π. The <u>vertical forms</u> for π are then those that are linear combinations of the one-forms

$$dx^i \quad .$$

Let (z^a), $1 \leq a,b \leq m$, be a set of local coordinates for Z such that

$$(x^i, z^a)$$

is a coordinate system for Z. (The local existence of such coordinates is a matter of the Implicit Function Theorem.) It is easy to see that the condition that $F^1(Z)$ be a direct sum (as modules) of $\mathcal{V}$ and E means that there are one-forms θ^a of the form

$$\theta^a = dz^a - f^a_i(x,z)dx^i \tag{5.1}$$

which form a basis for E. The θ^a (which are uniquely defined by the choice of coordinates) are called the <u>connection forms</u>.

The curves satisfying the differential equations

$$\frac{dz}{dt} = f(x,z) \tag{5.2}$$

are--in coordinate-free terms--the maps

$$\sigma: T \to Z$$

(T = interval of real numbers) such that:

$$\sigma^*(E) = 0 \tag{5.3}$$

Curves satisfying (5.3) are said to be <u>horizontal</u>. They are uniquely determined by their projection $t \to \pi\sigma(t)$ in X <u>and</u> their initial value $\sigma(0)$. As explained in IM, Vol. IX, they determine the geometric operation of "parallel transport" and "holonomy". For each curve $t \to \alpha(t)$, $a \leq t \leq b$, in X there is (modulo the "global" property of "horizontal completeness", which we assume for simplicity) a diffeomorphism.

$$\phi_\alpha: Z(\alpha(a)) \to Z(\alpha(b))$$

resulting from horizontal-lifting α to curves which start at various points of $Z(\alpha(a))$.

<u>Definition</u>. Let (Z,X,π), (Y,X,π') be fiber spaces, with the second a bundle of groups, acting on the first fiber space. A <u>Cartan-Ehresmann connection</u> for this set-up is an Ehresmann connection for (Z,X,π) such that the parallel-transport diffeomorphisms between fibers is a transformation group isomorphism. Explicitly, if

$$t \to \alpha(t) \quad , \qquad a \leq t \leq b \quad ,$$

is a curve in X, if

$$\phi_\alpha : Z(\alpha(a)) \to Z(\alpha(b))$$

is the diffeomorphism resulting from parallel-transporting along α, and if $g \in Y(\alpha(a))$, then there is a $g' \to Y(\alpha(b))$ such that:

$$\phi_\alpha(g(z)) = g'\phi_\alpha(z)$$
$$\text{for } z \in Z(\alpha(a)) \quad .$$

Remark. Here is what a "transformation group isomorphism" means in some detail. Let G,G' be groups which act as transformation groups on spaces S and S'. A map

$$\phi: S \to S'$$

is a "transformation group isomorphism" if ϕ^{-1} exists and if ϕ intertwines the action of G and G' on S and S', i.e., if $g \in G$, then

$$\phi g \phi^{-1} \in G' \quad .$$

For geometries defined in the "Felix Klein" way, via their groups of automorphisms, this typically means that ϕ is an isomorphism of the geometric structure.

The analytic conditions that a given Ehresmann connection have this additional "Cartan-Ehresmann" property are readily written down in terms of a local-product structure. This material was also presented in detail in IM, Vol. X. Rather than to on further with this "generalized nonsense" associated with the theory of connections (which I find rather boring!) I want to return now to the study of connections defined by prolongations and Bäcklund transformations.

Chapter 6

THE KORTEWEG-DE VRIES SOLITONS

1. INTRODUCTION

The work we have done for the Sine-Gordon equation is not at all special--it can be carried out for any equation which fits into the Estabrook-Wahlquist scheme [1]. In this chapter, I will develop material for the Korteweg-de Vries equation. In some respects the situation here is simpler than for Sine-Gordon.

2. THE BASIC KORTEWEG-DE VRIES PROLONGATION

Our basic reference is [1]. Consider independent variables (x,t). The Korteweg-de Vries equation is

$$u_t + u_{xxx} + 12\,uu_x = 0 \qquad (2.1)$$

Let X be the space of variables

$$(x,t,u,u_x,u_{xx}) \quad .$$

Let Y be the space of variables

$$(x,t,u,u_x,u_{xx},y) \quad .$$

The map $(x,t,u,u_x,u_{xx},y) \to (x,t,u,u_x,u_{xx})$ defines Y as a fiber space over X with one-dimensional fiber. Set:

$$= \quad dy + (2u + y^2 - \lambda)dx$$

$$- 4[(u+ \ )(2u + y^2 - \lambda) + \tfrac{1}{2}u_{xx} - u_x y]dt \qquad (2.2)$$

The basic result of [1] can be described as follows. If $u(x,t)$ is a solution of (2.1), the form θ' resulting from substituting it into θ is completely integrable. The "integral", i.e., the function $y(x,t)$ which makes $\theta' = 0$, generates another solution $u'(x,t)$ of (2.1) via the formula:

$$u' = -u - y^2 + \lambda \qquad (2.3)$$

(u,u') are related via the <u>Bäcklund transformation</u> [2].

The question we deal with here is: How does one find $y(x,t)$ and what are its group theoretic and geometric properties. In fact, as we have seen in previous work, θ defines a $SL(2,R)$-connection for the fiber space (2.2), and our plan is to relate the properties of this connection to the properties of the pair (u,u') of solutions of Korteweg-de Vries.

3. SL(2,R) AND QUADRATIC VECTOR FIELDS ON THE LINE

In this section we shall do some "pure" Lie theoretic calculations which are useful in analyzing the prolongation structure for Korteweg-de Vries (and all other equations whose "structure group" is $SL(2,R)$, for that matter).

Let $\mathcal{G}$ be the Lie algebra of vector fields in one real variable of the form

$$A = (a + by + cy^2)\frac{\partial}{\partial y} \quad . \tag{3.1}$$

We know this Lie algebra arises via SL(2,R) acting as linear fractional transformations. Our aim here is to make this as explicit as possible in order to be of maximal use in the study of prolongations.

First, introduce a two-dimensional linear space with variables z^i, $i = 1,2$, and the summation convention. Consider a linear vector field on this space:

$$A = a^i_j z^j \frac{\partial}{\partial z^i} \tag{3.2}$$

characterized by the 2×2 matrix

$$a = \begin{pmatrix} a^1_1 & a^1_2 \\ a^2_2 & a^2_2 \end{pmatrix}\begin{pmatrix} y^1 \\ y^2 \end{pmatrix} \tag{3.3}$$

Suppose it has trace zero, i.e.,

$$a^1_1 + a^2_2 = 0 \tag{3.4}$$

Set:

$$y = z^1/z^2 \tag{3.5}$$

Then,

$$A(y) = \frac{A(z^1)}{z^2} - \frac{z^1}{(z^2)^2} A(z^2)$$

$$= \frac{(a_1^1 z^1 + a_2^1 z^2)}{z^2} - \frac{z^1}{(z^2)^2} (z_1^2 z^1 + a_2^2 z^2)$$

$$= a_1^1 y + a_2^1 - a_1^2 y^2 - a_2^2 y \tag{3.6}$$

We can now compare (3.2) with (3.1). Notice that (3.6) says that A generates--in terms of the "inhomogenous" variables y given by (3.5),--a vector field of type (3.1), with the following relations between the coefficients:

$$a = a_2^1 \; ; \qquad b = a_1^1 - a_2^2 = 2a_1^1 \; ;$$
$$c = - a_1^2 \tag{3.7}$$

We can use these formulas to find the orbit curves of the vector field (3.1), i.e., the solutions $y(s)$ of

$$\frac{dy}{ds} = (a + by + cy^2) \quad . \tag{3.8}$$

Namely, the orbit curves of A acting on the space of variables z^1, z^2 are:

$$\frac{dz^i}{ds} = a_j^i z^j \quad ,$$

or

$$\frac{d\underline{z}}{ds} = \underline{a}\,\underline{z} \quad ,$$

where

$$\underline{z} = \begin{pmatrix} z^1 \\ z^2 \end{pmatrix}$$

Hence,

$$\underline{z}(s) = \exp(s\underline{a})\underline{z}(0) \quad . \tag{3.9}$$

Suppose that

$$\exp(s\underline{a}) = \begin{pmatrix} g_1^1(s), & g_2^1(s) \\ g_1^2(s), & g_2^2(s) \end{pmatrix} \tag{3.10}$$

Then, $y(s)$, the solution of (3.8), is of the following form:

$$y(s) = \frac{z^1(s)}{z^2(s)}$$

$$= \frac{g_1^1(s)z^2(0) + g_2^1(s)z^2(0)}{g_1^2(s)z^1(0) + g_2^2(s)z^2(0)}$$

$$= \frac{g_1^1(s)y(0) + g_2^1(s)}{g_1^2(s)y(0) + g_2^2(s)} \tag{3.11}$$

This exhibits the solution of (3.8) as a linear <u>fractional transformation</u>.

Of course, finding the coefficients in (3.11) explicitly depends on being able to exponentiate the matrix $\underline{a}$. This is readily accomplished if $\underline{a}$ can be put into triangular form, for example.

The formulas also hold if (3.8) depends on s explicitly on the right hand side, i.e., if

$$\frac{dy}{ds} = a(s) + b(s)y + c(s)y^2 \tag{3.12}$$

Now, set

$$\underline{a}(s) = \begin{pmatrix} a_1^1(s) & a_2^1(s) \\ a_1^2(s) & a_2^2(s) \end{pmatrix}$$

with formulas (3.7) defining the coefficients of the matrix in terms of a,b,c as given in (3.12). Then, (3.11) continues to hold, where

$$\begin{aligned} \frac{d\underline{g}}{ds} &= \underline{a}(s)\underline{g}(s) \\ \underline{g}(0) &= 1 \quad . \end{aligned} \tag{3.13}$$

4. SOLUTIONS OF KORTEWEG-DE VRIES AS SURFACES IN SL(2,R)

Return to the situation of Section 2. Suppose $u(x,t)$ is a solution of Korteweg-de Vries. $y(x,t)$ is then a solution of:

$$dy = (2u + y^2 - \lambda)dx - 4[(u+\lambda)(2u + y^2 - \lambda) + \tfrac{1}{2}u_{xx} - yu_x]dt \tag{4.1}$$

We know that this Pfaffian equation is <u>completely integrable</u>. Thus, $y(x,t)$ can be expressed in terms of $y(0,0)$. In fact, this is of the following form:

$$y(x,t) = \frac{g_1^1(x,t)y(0,0) + g_2^1(x,t)}{g_1^2(x,t)y(0,0) + g_2^2(x,t)} \quad . \tag{4.2}$$

$$(x,t) \rightarrow \underline{g}(x,t) = \begin{pmatrix} g_1^1(x,t), & g_1^2 \\ g_2^1 & g_2^2 \end{pmatrix} \tag{4.3}$$

is a map of $R^2 \rightarrow SL(2,R)$ which <u>depends on</u> u. In fact, if we require that

$$\underline{g}(0,0) = 1 \quad , \quad \text{the identity element} \quad ,$$

then this map is uniquely determined by u, and identifies it with a "surface" in the three-parameter space $SL(2,R)$.

Of course, finding $\underline{g}(x,t)$ explicitly requires solving differential equations. Here is one relatively easy way to do it.

Introduce a new constant α, and set

$$t = \alpha x \quad . \tag{4.4}$$

Set:

$$u(s) = u(s,\alpha s) \tag{4.5}$$

$$y(s) = y(s,\alpha s) \tag{4.6}$$

For the moment, suppose u and y are functions of the

single real variable s. Then (4.1) takes the form

$$\frac{dy}{ds} = (2u(s) - \lambda - 4\alpha\lambda(2u-\lambda) + \tfrac{1}{2}u_{xx}) + 4\alpha u_x y + (1 - 4\alpha\lambda]y^2 \tag{4.7}$$

Thus, we set (following (3.6)

$$\underline{a} = \begin{pmatrix} 2\alpha u_x , & 2u - \lambda - 4\alpha\lambda(2u-\lambda) + \tfrac{1}{2}u_{xx} \\ 4\alpha\lambda - 1, & - 2\alpha u_x \end{pmatrix} \tag{4.8}$$

Thus,

$$\frac{d\underline{g}}{ds} = \underline{a}\ \underline{g} , \qquad \underline{g}(0) = 1 .$$

$\underline{g}$ is really a function of $\underline{g}(s,\alpha)$. We see that

$$x = s ;$$

$$t = \alpha x .$$

Thus,

$$(x,t) \to \underline{g}\left(x, \frac{x}{t}\right)$$

is the surface in $SL(2,R)$ determined by $u(x,t)$.

Of course, finding $\underline{g}$ cannot be done in _general_ "by quadratures". However, these equations can be solved for the _solitons_. This is now our goal.

5. THE REST SOLUTION OF KORTEWEG-DE VRIES AS A SURFACE IN SL(2,R)

The "solitons" are (for our purposes) defined as the solutions of Korteweg-de Vries which are generated from the "rest" solution $u \equiv 0$ by iterated Bäcklund transformations. (They come with natural parameters because Bäcklund transformations are not "mappings" by "correspondences". Thus, the "image" of u is a u' depending on a parameter. There is also the "additional" parameter λ available.)

With this simplifying choice, the one-form θ takes the following form:

$$\theta = dy + (y^2-\lambda)dx - 4\lambda(y^2-\lambda)dt \tag{5.1}$$

Set:

$$s = x - 4\lambda t \tag{5.2}$$

Then, θ can be written as:

$$\theta = dy + (y^2-\lambda)ds \tag{5.3}$$

We can now find the surface in SL(2,R) corresponding to the solution $u \equiv 0$ of Korteweg-de Vries by solving $\theta = 0$ in terms of linear fractional transformations. Of course, we could do this by using the general formulas developed in Section 4. However, it is just as easy to work directly with (5.3).

$y(s)$ should satisfy

$$\frac{dy}{ds} = y^2 - \lambda \quad , \tag{5.4}$$

$$\frac{dy}{y^2-\lambda} = ds \quad ,$$

$$\frac{1}{y^2-\lambda} = \frac{1}{\sqrt{\lambda}}\left(\frac{\frac{1}{2}}{y - \sqrt{\lambda}} - \frac{\frac{1}{2}}{y + \sqrt{\lambda}}\right)$$

Hence, integrating (5.4) we have:

$$\log \frac{y-\sqrt{\lambda}}{y+\sqrt{\lambda}} = 2\sqrt{\lambda}\, s \quad ,$$

or

$$\frac{y-\sqrt{\lambda}}{y+\sqrt{\lambda}} = ce^{2\sqrt{\lambda}s}$$

$$y - \sqrt{\lambda} = (y+\sqrt{\lambda})ce^{2\sqrt{\lambda}s}$$

$$y(1-ce^{2\sqrt{\lambda}s}) = \sqrt{\lambda}ce^{2\sqrt{\lambda}s} + \sqrt{\lambda}$$

or

$$y(s) = \sqrt{\lambda}\left(\frac{ce^{2\sqrt{\lambda}s} + 1}{1 - ce^{2\sqrt{\lambda}s}}\right) \tag{5.5}$$

$$c = \frac{y(0) - \sqrt{\lambda}}{y(0) + \sqrt{\lambda}} \tag{5.6}$$

$$y(s) = \sqrt{\lambda}\left(\frac{1 + \left(\dfrac{y(0) - \sqrt{\lambda}}{y(0) + \sqrt{\lambda}}\right) e^{2\sqrt{\lambda}s}}{1 - \left(\dfrac{y(0) - \sqrt{\lambda}}{y(0) + \sqrt{\lambda}}\right) e^{2\sqrt{\lambda}s}}\right)$$

$$= \sqrt{\lambda}\ \frac{(y(0) + \sqrt{\lambda} + e^{2\sqrt{\lambda}s}\ (y(0) - \sqrt{\lambda}))}{y(0) + \sqrt{\lambda} - e^{2\sqrt{\lambda}s}\ (y(0) - \sqrt{\lambda})}$$

$$= \frac{y(0)(1 + e^{2\sqrt{\lambda}s}) + \sqrt{\lambda}\ (1 - e^{2\sqrt{\lambda}s})}{\dfrac{y(0)}{\sqrt{\lambda}}\ (1 - e^{2\sqrt{\lambda}s}) + (1 + e^{2\sqrt{\lambda}s})} \tag{5.7}$$

$=$, from (3.11),

$$\frac{g_1^1(s)y(0) + g_2^1}{g_1^2 y(0) + g_2^2} \tag{5.8}$$

We can now compare (5.7) with (5.8) to obtain the matrix $\underline{g}(s)$

$$\begin{aligned} g_1^1(s) &= 1 + e^{2\sqrt{\lambda}s} \\ g_2^1 &= \sqrt{\lambda}\ (1 - e^{2\sqrt{\lambda}s}) \\ g_1^2 &= \frac{1 - e^{2\sqrt{\lambda}s}}{\sqrt{\lambda}} \\ g_2^2 &= 1 + e^{2\sqrt{\lambda}s} \end{aligned} \tag{5.9}$$

Of course, this method does not guarantee us a matrix of determinant +1; to do this, the g's must be rescaled.

$$\det \underline{g} = (1 + e^{2\sqrt{\lambda}s})^2 - (1 - e^{2\sqrt{\lambda}s})^2$$

$$= 4e^{2\sqrt{\lambda}s}$$

Hence, everything must be rescaled by dividing by $2e^{\sqrt{\lambda}s}$. After this is done we obtain:

$$\begin{aligned} g_1^1(s) &= \cosh \sqrt{\lambda}s \\ g_2^1 &= \sqrt{\lambda} \sinh \sqrt{\lambda}s \\ g_1^2 &= \sqrt{\lambda} \sinh -\sqrt{\lambda}s \\ g_{22} &= \cosh \sqrt{\lambda}s \end{aligned} \tag{5.10}$$

These are familiar formulas in Soliton Theory (and in Lie theory also).

We can sum up as follows.

Theorem 5.1. To the rest solution $u = 0$ if the Korteweg-de Vries equation belongs the surface

$$(x,t) \to \underline{g}(x - 4\lambda t) \tag{5.11}$$

In $SL(2,R)$, where $\underline{g}(s)$ is the matrix given by formulas (5.10).

Note that

$$s \to \underline{g}(s)$$

is a one-parameter subgroup of $SL(2,R)$. If:

$\lambda > 0$, it is a non-compact one parameter group

$\lambda < 0$, it is a <u>compact</u> group

6. THE PROLONGATION EQUATIONS FOR SIMILARITY SOLUTIONS OF KORTEWEG-DE VRIES

Consider the Pfaffian equation:

$$\begin{aligned}\omega \;=\; & dy + (2u + y^2 - \lambda)dx \\ & - 4[(u+\lambda)(2u + y^2 - \lambda) + \tfrac{1}{2}u_{xx} - u_x y]dt\end{aligned} \tag{6.1}$$

We know that if $u(x,t)$ satisfies the Korteweg-de Vries equation, then ω is <u>completely integrable</u>, i.e., there exist functions $y(x,t)$ which when substituted into (6.1) make ω identically zero. However, finding these functions involves solving a Riccati differential equation, which is (unless it has constant coefficients) impossible to do in general.

However, if u is a "one-soliton" solution of Korteweg-de Vries, it is a so called <u>similarity solution</u> of Korteweg-de Vries, i.e., is of the form

$$u(x,t) = v(s) \quad , \tag{6.2}$$

with

$$s = x - at \tag{6.3}$$

We shall now investigate the simplifications this presents in the problem of finding $y(x,t)$.

Set

$$s = c = \text{constant} \quad . \tag{6.4}$$

Then,

$$dx = a\,dt \quad . \tag{6.5}$$

Restrict ω to the submanifold determined by condition (6.4); denote the resulting form by ω_c. Now u, u_x, u_{xx} are constant on these submanifolds. Denote their values as $u(c), u_x(c), u_{xx}(c)$. Thus, we have:

$$\begin{aligned} \omega_c &= dy + (2u + y^2 - \lambda)a\,dt \\ &\quad - 4[(u+\lambda)(2u + y^2 - \lambda) + \tfrac{1}{2}u_{xx} - u_x y]dt \\ &= dy + (2au - 8(u+\lambda)u + \tfrac{1}{2}u_{xx})dt \\ &\quad - u_x y\,dt \\ &\quad + (a - 4(u+\lambda))(y^2-\lambda)dt \end{aligned} \tag{6.5}$$

Thus, if we look for y as a function of t (with c as a "parameter") such that the right hand side of (6.5) is zero, we have to solve a Riccati differential equation with constant coefficients. This immediately tells us what the analytical form of g must be.

Theorem 6.1. Let

$$\underline{g}(\alpha,\beta,\gamma,t) = \begin{pmatrix} g_1^1(\alpha,\beta,\gamma,t) \ , & g_2^1 \\ g_1^2 & g_2^2 \end{pmatrix}$$

be the general solution of the Riccati equation:

$$\frac{dz}{dt} = \alpha + \beta z + c(z^2 - \ )$$

i.e.,

$$z(t) = \frac{g_1^1 z(0) + g_2^1}{g_1^2 z(0) + g_2^2} \qquad (6.6)$$

for all t .

Let $y(x,t)$ be a function which solves the Pfaffian equation $\omega = 0$. Then,

$$y(at + c,t) = \frac{g_1^1(\alpha,\beta,\gamma,t)y(c,0) + g_2^1}{g_1^2 y(x,0) + g_2^2} \qquad (6.7)$$

with

$$\alpha = -2au(c) + 8(u(c)+\lambda)u(c) + \tfrac{1}{2}u_{xx}(c) \qquad (6.8)$$

$$\beta = u_x(c) \qquad (6.9)$$

$$\gamma = 4(u(c)+\lambda) - a \qquad (6.10)$$

To determine y it only remains to find

$$y(x,0) \equiv z(x) \qquad (6.11)$$

This can be done by setting $dt = 0$ in (6.1), and making the substitution (6.6). The result is:

$$\frac{dz}{dx} = 2v(x) + y^2 - \lambda \quad , \tag{6.12}$$

with

$$v(x) = u(x,0) \quad . \tag{6.13}$$

This is again a Riccati differential equation, but with non-constant coefficients. Let us sum up as follows:

Theorem 6.2. Let $y(x,t)$ be a solution of the completely integrable Pfaffian equation $\omega = 0$ with $u(x,t)$ a solution of the Korteweg-de Vries equation of the form

$$u(x,t) = v(x-at) \quad .$$

Let G denote the vector group $SL(2,r)$, considered as a transformation group on R via linear fractional transformations. Then, there are maps $(x,t) \to g_1(x,t)$, $x \to g_2(x)$ of $R^2 \to G$ and $R \to G$ beginning at the identity element for $x = t = 0$, such that

$$y(x,t) = g_1(x,t) g_2(x) y(0,0) \quad .$$

The maps are determined by the data as explained above.

Of course, this algorithm for writing down explicitly the solution $y(x,t)$ flounders in the fact that in general we do not know anything more about $v(x)$, so (6.12) is a non-constant coefficient Riccati equation of the usual contractable type. However, we might be interested in the case that $u(x,t)$ is itself associated with the "one-soliton" solution to Korteweg-de Vries. This means that v itself will satisfy a differential equation. We shall investigate this situation in a later section.

Chapter 7

THE GLOBAL STRUCTURE OF THE KORTEWEG-DE VRIES SOLITON HEIRARCHY

1. INTRODUCTION

The "solitons" of the Korteweg-de Vries equation (and the other equations that are covered by the Estabrook-Wahlquist ideas) are cross-sections of bundles with one-dimensional fibers and with SL(2,R) as structure group. The "kinks" are interesting global invariants, developed in the work of D. Finkelstein, Misner, Perring, Skyrme and Rubenstein. An obvious program is to relate the "kinks" to global invariants of the bundle and the connections defined on them. Having done this, we will be in a good position to extend the ideas even further. In order to fix the ideas, I will now concentrate on the Korteweg-de Vries equation.

2. THE KORTEWEG-DE VRIES BÄCKLUND CONNECTION

We shall now describe the Korteweg-de Vries Bäcklund transformation [1,2] as a connection for an SL(2,R)-bundle. Consider the following system of first order partial differential equations:

$$w'_x + w_x = (w' - w)^2 - \lambda \qquad (2.1)$$

$$w'_t + w_t = 4((u')^2 + u'u + u^2) + 2(w - w')(u'_x - u_x) \qquad (2.2)$$

$$u = -w_x \qquad (2.3)$$

$$u' = -w'_x$$

Let $u(x,t)$ be a solution of the Korteweg-de Vries equation:

$$u_t + u_{xxx} + 12\,uu_x \qquad (2.4)$$

Let $w(x,t)$ be a function such that

$$u(x,t) = -w_x \quad . \qquad (2.5)$$

Let $y(x,t)$ be a "pseudopotential" for u, i.e., y satisfies the following Pfaffian equation:

$$dy = (\lambda - y^2 - 2u)dx + 4[(u+\lambda)(2u+y^2-\lambda) + \tfrac{1}{2}u_{xx} - u_x y]dt \qquad (2.6)$$

(λ is a parameter that is fixed for the moment.) Now, set:

$$w' = w - y \qquad (2.7)$$

Then,

$$dw' = (w_x - \lambda + (w'-w)2 + 2u)dx + 4[w_t - (u+\lambda)(2u + (w'-w)^2 - \lambda) - \tfrac{1}{2}u_{xx} + u_x(w-w')]dt \qquad (2.8)$$

This can be interpreted as a connection equation. Set

$$\theta = dw' - (\text{right hand side of } (2.8)) \quad .$$

θ defines a <u>connection</u> for the fiber space whose basis is the space of variables $x,t,u,u_x,\ldots,$ (since using (2.1) w_t can be expressed in terms of derivatives of x) and whose fiber variable is w'. As such, it is an $SL(2,R)$ <u>connection</u> (with $SL(2,R)$ acting on the fiber via linear fractional transformation). It will be called the <u>Korteweg-de Vries Bäcklund connection</u>.

We recognize that $w'(x,t)$ defines the <u>Bäcklund transform</u> of $u(x,t)$. My plan is to set

$$dt = 0 \quad ,$$

and study the resulting bundle and connection from the <u>global</u> point of view.

3. THE KORTEWEG-DE VRIES HEIRARCHY

Consider a sequence

$$w_n(x,t) \quad , \qquad n \geq 0, \lambda_n$$

of real-valued functions on R^2 and constants λ_n. Set:

$$u_n = -(w_n)_x \quad . \tag{3.1}$$

Suppose they satisfy the following conditions:

$$w_0 = 0 \tag{3.2}$$

$$\theta_{n+1} = dw_{n+1} - (-\lambda_n + (w_{n+1} - w_n)^2 - (w_n)_x)dx$$
$$- 4[(w_n)_t - (u_n+\lambda_n)(2u_n + (w_{n+1} - w_n)^2 - \lambda)$$
$$- \tfrac{1}{2}(u_n)_{xx} + (u_n)_x(w_n - w_{n+1})]dt$$
$$= 0 \qquad (3.3)$$

For each n, we see that the one-form θ_{n+1} defines an SL(2,R)-connection on the fiber space whose fiber coordinate is w_{n+1}. The sequence (w_n,λ_n) of fucntions is called the Korteweg-de Vries Soliton Heirarchy. The sequence of bundles and connections θ_n is called the Soliton connection heirarchy.

Think of the physical significance of the $w_n(x,t)$ as "waves". "x" is the space parameter. We are intensely interested in the function

$$x \to w_n(x,t) \qquad , \qquad (3.4)$$

with t held fixed. Since Equations (3.3) are completely integrable, the $w_n(x,t)$ for all t is determined by the behavior of the function (3.4) at one value of t. Thus, it is a reasonable project to freeze t at one value--which might as well be $t = 0$ --and consider the connections that result. (Luckily, this also simplifies the mathematics!) With this done, θ_{n+1} becomes:

$$\theta_{n+1} = dw_{n+1} + (\lambda_n - (w_{n+1} - w_n)^2 + (w_n)_x)dx \qquad (3.4)$$

We now want to know if this can be interpreted as a connection for a fiber space with a compact base and fiber. This requires compacting the x- and w-spaces, which, of course, may be done by adding the "point at infinity". Let us begin with the case $n = 0$.

4. THE BUNDLE FOR THE ONE-SOLITONS

Set $n = 0$ with

$$w_0 = 0 \quad .$$

Then

$$\theta_1 = dw_1 + (\lambda_0 - w_1^2)dx \tag{4.1}$$

Let us compactify the space of variables (x,w_1) by "Cayley transforming" them into the product $S_1 \times S_1$ of the unit circle of $\mathbb{C}$ with itself:

$$z_1 = \frac{w_1 + i}{w_1 - i} \tag{4.2}$$

$$s = \frac{x + i}{x - i} \tag{4.3}$$

$$\theta = \frac{dz_1}{(z_1-1)^2} + \left(\lambda_0 + i\left(\frac{z_1+1}{z_1-1}\right)^2\right)\frac{ds}{(s-1)^2}$$

or

$$(s-1)^2(z_1-1)^2\theta_1 = (s-1)^2 dz_1 + (\lambda_0(z_1-1)^2 + (z_1+1)^2)ds \tag{4.4}$$

We can read off the global story from this formula. Let

$$Y = \{(z_1,s) \in \mathbb{C}^2: |z_1| = |s| = 1\}$$

$$X = \{(z_1) \in \mathbb{C}: |z_1| = 1\}$$

$$\pi: Y \to X \quad \text{the map} \quad \pi(z_1,s) = z_1 \quad .$$

π defines Y as a fiber space with S_1 as base and S_1 as fiber (in fact, it is a "product"). $SL(2,R)$ ($\equiv SU(1,1)$) is the "structure group", in the sense that it acts on each fiber.

Consider the one-form

$$\theta_1' = (s-1)^2 dz_1 + (\lambda_0(z_1-1)^2 + (z_1+1)^2)ds \tag{4.5}$$

It defines a "connection" for the fiber space, but with a "singularity" on the fiber above the point $s = 1$.

This has been a warm-up. Let us now turn to the more interesting case of the two-soliton fiber space.

5. THE TWO-SOLITON FIBER SPACE

Specialize (3.4) to $n = 1$:

$$\theta_2 = dw_2 + (\lambda_1 - (w_2-w_1)^2 + (w_1)_x)dx \tag{5.1}$$

Also,

$$dw_1 = -(\lambda_0 - w_1^2)dx \quad ,$$

i.e.,

$$(w_1)_x = w_1^2 - \lambda_0 \tag{5.2}$$

Hence,

$$\theta_2 = dw_2 + (\lambda_1 - \lambda_0 - w_2^2 - w_1^2 = 2w_2w_1 + w_1^2)dx$$

$$= dw_2 + (\lambda_1 - \lambda_0 - w_2^2 + 2w_2w_1)dx \tag{5.3}$$

Now let us change variables as in (4.4)-(4.5):

$$w_2 = i\left(\frac{z_2 + 1}{z_2 - 1}\right)$$

Using (4.6)-(4.7), we have:

$$\theta_2 = \frac{dz_2}{(z_2-1)^2} + \left(\lambda_1 - \lambda_0 + \frac{(z_2+1)^2}{(z_2-1)^2} - 2\,\frac{(z_2+1)}{(z_2-1)}\,\frac{(z_1+1)}{(z_1-1)}\right)\frac{ds}{(s-1)^2}$$

Now, set:

$$\theta_2' = (s-1)^2(z_2-1)^2(z_1-1)\theta_2$$

$$\begin{aligned} = (s-1)^2(z_1-1)dz_2 + [&(\lambda_1-\lambda_0)(s-1)^2(z_2-1)^2(z_1-1) \\ &+ (z_2+1)^2(s-1)^2(z_1-1) \\ &- 2(z_2^2-1)(z_1+1)]ds \end{aligned}$$

This formula exhibits θ_2' as a one-form which is everywhere defined on $S_1 \times S_1 \times S_1$, with a "singularity", as a <u>connection</u>, on the submanifold of the bases $s = 1 = z_1$. This construction can clearly be continued to the rest of the soliton heirarchy. Thus, we construct a tower of product bundles

$$\cdots \to Y_2 \to Y_1 \to X \quad ,$$

with fibers all S_1, with forms globally defined in these bundles, but having singularities as <u>connections</u>.

<u>Remark</u>. One might think that an alternate geometric route is open: Define the fiber spaces as "twisted" so that the "twists" precisely compensate for the singularity of the connection at $s = 1$. In this way, one might hope to construct non-product $SL(2,R)$-bundles with <u>globally defined</u>, non-singular connections. In fact, I have not been able--after considerable effort--to construct them, and I suspect that they do not exist. I believe this is related to the "non-Fuchsian" behavior of the "inverse scattering" equations

$$\psi_{xx} + u\psi = \lambda\psi$$

at $x = \infty$. I plan further research in this direction.

BIBLIOGRAPHY

1. H.D. Wahlquist and F.B. Estabrook, "Prolongation Structures of Nonlinear Evolution Equations", <u>J. Math. Phys</u>. <u>16</u> (1975). pp. 1-7.

2. H.D. Wahlquist and F.B. Estabrook, <u>Phys. Rev. Lett</u>. <u>31</u> (1973), 1386.

Chapter 8

THE KORTEWEG-DE VRIES SOLITON HEIRARCHY AS "KINKS"

1. INTRODUCTION

The "kink" concept is one of the most interesting mechanisms for introducing "topological" ideas into physics. It also has great mathematical interest as an essential mathematical feature of the theory of nonlinear partial differential equations. In previous chapters I have briefly discussed its properties for "Sine-Gordon" solitons. (H. Morris [1] has developed a more definitive theory using another approach.) I now want to examine the same material in the context of the Korteweg-de Vries equation.

2. THE EQUATIONS FOR THE POTENTIALS OF THE KORTEWEG-DE VRIES SOLITON HEIRARCHY

Let $\mathcal{W}$ be a set of continuous real-valued functions $w(x)$, $-\infty < x < \infty$. In order to talk about a "kink" structure for elements of $\mathcal{W}$ we must be given the following data: a mapping $\phi\colon R \to S_1$ ($\equiv$ unit circle in R^2) such that, for each $w \in \mathcal{W}$, the limits

$$\lim_{x\to\pm\infty} \phi(w(x)) = p_{\pm\infty}(w)$$

exist. If this condition is satisfied, each $w \in \mathcal{W}$ determines the continuous curve

$$x \to \phi(w(x))$$

going from $p_{-\infty}(w)$ to $p_{+\infty}(w)$ on S_1; this curve belongs to a homotopy class of such curves, which is identified with an integer in the standard way. ($\pi_1(S_1) = X$.) This integer is called the kink of w.

In this section I will describe a class of curves for which this "kink" number seems to be well defined.

As in the previous chapter, define the concept of a Korteweg-de Vries soliton heirarchy:

$$(u_n, y_n, \lambda_n) \quad , \qquad n = 0,1,\ldots$$

Now, these functions depend on two variables x and t. Fix t, obtaining functions that we also denote by u_n, y_n. (The reader should keep in mind that the variable "t" is fixed throughout this chapter with no special notational distinction.) The equations presented in the previous chapter then specialize as follows:

$$\frac{dy_n}{dx} = \lambda_n - y_n^2 - 2u_n \tag{2.1}$$

$$u_{n+1} = -u_n - y_n^2 + \lambda_n \tag{2.2}$$

for $n \geq 0$.

$$u_0 = 0 \quad . \tag{2.3}$$

In order to put these equations into a more convenient form, introduce (along with Estabrook and Wahlquist) the "potential" function w_n such that

$$\frac{dw_n}{dx} = u_n \tag{2.4}$$

Then, combining (2.2) and (2.4):

$$\frac{d}{dx}(w_{n+1} + w_n) = \lambda_n - y_n^2$$

$$= \frac{dy_n}{dx} + 2u_n$$

$$= \frac{dy_n}{dx} + 2\frac{dw_n}{dx} \quad ,$$

or

$$w_{n+1} = w_n + y_n + (\text{constant}) \quad .$$

The arbitrary constant involved in the definition (2.4) of w_n in terms of u_n now can be chosen recursively so that

$$w_{n+1} = w_n + y_n \quad . \tag{2.5}$$

Then, (2.1)-(2.2) and (2.5) determine the following equations:

$$\frac{dw_{n+1}}{dx} = \frac{dw_n}{dx} - (w_{n+1} - w_n)^2 + \lambda_n - 2\frac{dw_n}{dx} \quad ,$$

or

$$\frac{dw_{n+1}}{dx} = -\frac{dw_n}{dx} - (w_{n+1} - w_n)^2 + \lambda_n \tag{2.6}$$

$$w_0 = 1 \quad . \tag{2.7}$$

Notice that Equations (2.6)-(2.7) --which are the definitive ones (and which are the defining equations for the Bäcklund transformation)--form a Riccati equation for w_{n+1}, in terms of w_n. They define ordinary differential equations of the following form

$$\frac{dw_{n+1}}{dx} = f_{n+1}(w_{n+1}, w_n, \ldots, w_1) \quad , \tag{2.8}$$

with $\lambda_n, \ldots, \lambda_0$ as parameters.

3. THE DEFINITION OF THE KINK NUMBER FOR KORTEWEG-DE VRIES

This development brings out even more clearly the analogy between Sine-Gordon and Korteweg-de Vries, and enables us to define the kink number in a reasonable way.

Let $(x,t) \to u(x,t)$ be a solution of Korteweg-de Vries. Fix t, and define the map

$$x \to \frac{\int_0^x u(x,t) + i}{\int_0^x u(x,t) - i} \equiv \alpha(x) \tag{3.1}$$

of $R \to S_1$. (Identify S_1 with the complex numbers of absolute value one.) Assume $\alpha(\pm\infty)$ exist. Then, the kink is the homotopy class determined by this curve. If $\alpha(+\infty) = \alpha(-\infty)$, then this integer is

$$\frac{1}{i} \int_{-\infty}^{\infty} \frac{d\alpha(x)}{\alpha(x)} \quad . \tag{3.2}$$

4. DECOUPLING OF THE FIRST TWO TERMS OF THE KORTEWEG-DE VRIES SOLITON HEIRARCHY VIA THE WAHLQUIST-ESTABROOK SUPERPOSITION FORMULA

Let us now return to the material treated in Section 2. From now on, by the "Korteweg-de Vries heirarchy" we shall mean the equations (2.6)-(2.7). Consider them for $n = 1$:

$$\frac{dw_1}{dx} = -w_1^2 + \lambda_0 \tag{4.1}$$

$$\frac{dw_2}{dx} = -\frac{dw_1}{dx} - (w_2 - w_1)^2 + \lambda_1 \tag{4.2}$$

$$= w_1^2 - \lambda_0 - (w_2 - w_1)^2 + \lambda_1$$

$$= 2w_1w_2 - w_2^2 + {}_1 - {}_0$$

Equations (4.1)-(4.2) constitute a system of ordinary differential equations for w_1, w_2.

Let us see if they can be "decoupled". Set:

$$w_2 = c(w_1 - z)^{-1} \quad . \tag{4.3}$$

where $z(x)$ is a function of x which is to be determined.

c is a constant. (The form chosen is suggested by the "superposition principle" for Korteweg-de Vries found by Wahlquist and Estabrook.)

$$\frac{dw_2}{dx} = -c(w_1 - z)^2\left(-w_1^2 + \lambda_0 - \frac{dz}{dx}\right)$$

$$= \text{ , using (4.2) ,}$$

$$w_1^2 + (\lambda_1 - \lambda_0) - (c(w_1 - z - 1) - w_1)^2 \qquad (4.4)$$

Multiply (4.4) through by $(w_1 - z)^2$; the following relation is obtained:

$$-c\left(-w_1^2 + \lambda_0 - \frac{dz}{dx}\right) = (w_1^2 + (\lambda_1 - \lambda_0))(w_1 - z)^2 - (c - w_1(w_1 - z))^2$$

$$= w_1^2(w_1 - z)^2 + (\lambda_1 - \lambda_0)(w_1 - z)^2$$

$$- c^2 + 2cw_1(w_1 - z) - w_1^2(w_1 - z)^2$$

$$= (\lambda_1 - \lambda_0)(w_1 - z)^2 - c^2 + 2cw_1(w_1 - z)$$

$$= (\lambda_1 - \lambda_0)(w_1^2 - 2zw_1 + z^2) - c^2 + 2cw_1^2$$

$$- 2czw_1$$

$$= (\lambda_1 - \lambda_0 + 2c)w_1^2 - 2w_1 z(\lambda_1 - \lambda_0 + c)$$

$$+ (\lambda_1 - \lambda_0)z^2 - c^2$$

$$(4.5)$$

Now set:

$$c = \lambda_0 - \lambda_1 \quad . \tag{4.6}$$

Hence,

$$\lambda_1 - \lambda_0 + 2c = c \tag{4.7}$$

Equation (4.5) collapses to the following equation:

$$\frac{dz}{dx} = c\lambda_0 + (\lambda_1-\lambda_0)z^2 - c^2$$

$$= (\lambda_0-\lambda_1)\lambda_0 - (\lambda_0-\lambda_1)^2 + (\lambda_1-\lambda_0)z^2$$

$$= \lambda_0\lambda_1 - \lambda_1^2 + (\lambda_1-\lambda_0)z^2$$

$$= (\lambda_1-\lambda_0)(z^2-\lambda_1) \tag{4.8}$$

Equation (4.8)--which is equivalent to (4.2)--is a <u>constant coefficient Riccati equation which is completely decoupled from</u> w_1.

We can now write down the <u>general solution</u> of the equations (4.1) and (4.8):

$$w_1(x) = g_1(x)(w_1(0)) \tag{4.9}$$

$$z(x) = g_2(x)(z(0)) \quad , \tag{4.10}$$

where $x \to g_1(x),\ g_2(x)$ are one-parameter subgroups of $SL(2,R)$ which <u>commute</u>. ($SL(2,R)$ acts on real numbers via linear fractional transformations.) Hence, via (4.3),

$$w_2(x) = (\lambda_0-\lambda_1)(g_1(x)(w_1(0)) - g_2(x)(z(0))^{-1} \quad (4.11)$$

$$w_2(0) = (\lambda_0-\lambda_1)(w_1(0) - z(0))^{-1} \quad (4.12)$$

We can now readily determine the asymptotic behavior of $w_1(x)$ and $w_2(x)$ as $x \to \pm\infty$. First, w_1. It is determined by differential Equation (4.1). The fixed points of the vector field

$$(\lambda_0 - w_1^2)\frac{\partial}{\partial w_1}$$

occur at

$$\pm\sqrt{\lambda_0} \quad .$$

The linear variational equations are

$$\frac{d(\delta w_1)}{dx} = -2w_1(\delta w_1) = \mp 2\sqrt{\lambda_0}\,\delta w_1 \quad .$$

Thus, $-\sqrt{\lambda_0}$ is the attractor critical point, $+\sqrt{\lambda_0}$ is the repulsor. Hence,

$$\lim_{x\to\pm\infty} w_1(x) = \mp\sqrt{\lambda_0}$$

Similarly,

$$\lim_{x\to\pm\infty} z(x) = \pm\sqrt{\lambda_1} \qquad (\text{if } \lambda_1 > 0) \quad .$$

Hence,

$$\lim_{x\to\pm\infty} w_2(x) = \frac{\lambda_0 - \lambda_1}{(\mp\sqrt{\lambda_0} \mp \sqrt{\lambda_1})} = \pm(\sqrt{\lambda_0} - \sqrt{\lambda_1})$$

To determine the "kink" number we must determine for how many x's, $w_2(x)$ takes on a given value. Let us examine this for $w_2(x) = \infty$, i.e., the set of x's such that

$$g_1(x)(w_1(0)) = g_2(x)z(0) \quad ,$$

i.e.,

$$g_2(x)^{-1}g_1(x)w_1(0) = z(0) \tag{4.13}$$

Now, $x \to g_1(x)$, $g_2(x)$ are one parameter subgroups of $SL(2,R)$ which commute. Hence

$$x \to g_2(x)g_1(x)$$

is also a one-parameter subgroup. It is readily seen that there are (generally) two values of x which satisfy (4.13), i.e., the kink number is 2.

Thus we see that all the standard qualitative properties of the two-soliton solutions of Korteweg-de Vries are elementary consequences of the decoupling of the differential equations into those for one-solitons. This decoupling is, on the mathematical side, just the "superposition formula" of Estabrook and Wahlquist (which is also a generalization of the "theorem of permutability" of Bianchi for Sine-Gordon) and on the physical side is the property that the two-solitons are "hard states" of the one-solitons. Of course, the beauty of this situation is that the formulas describing this "superposition" are explicit and elementary.

Chapter 9

AN EXAMPLE OF A GAUGE TRANSFORMATION DESINGULARIZATION OF A CONNECTION VIA THE FUCHSIAN THEORY OF ORDINARY DIFFERENTIAL EQUATIONS

1. INTRODUCTION

The characteristic physical feature of connections is their law of transformation under <u>gauge transformation</u>. From the viewpoint of tensor analysis, this means that they do <u>not</u> transform "tensorially", i.e., their components can vanish in one "frame" but not in another. In particular, there is a possibility of singularities being only "apparent", i.e., being present in one "gauge", but being cancelled out by an appropriate "gauge" transformation. In particular, this possibility has played a strong conceptual role in recent speculations about "unified" field-elementary particle theories using "gauge" fields.

Now, this subject is not too well developed mathematically--there is no body of standard lore to fall back on. In this chapter I will give an example that I hope will illustrate the essence of what is involved.

Here is what is going on analytically. Suppose that X is a manifold, $X_0 \subset X$ a submanifold, Y a space with a Lie group G acting on Y. Set

$$E = X \times Y$$

Regard E as a fiber space, with X as base. A <u>connection</u>, with G as structure group, is then determined as a $\mathcal{G}$-valued one-form

$$\theta\colon T(X) \to \mathcal{G} \quad .$$

($\mathcal{G}$ = Lie algebra of G.) Suppose now that θ is not defined on all of X, but only on $X - X_0$. Intuitively, think of θ as developing some sort of "singularity" on the submanifold X_0.

Let $\phi\colon X - X_0 \to G$ be a map, i.e., a "gauge transformation". It defines a diffeomorphism

$$(X-X_0) \times Y \to (X-X_0) \times Y$$

as follows:

$$(x,y) \to (x, \phi(x)y) \tag{1.1}$$

The connection θ then transforms into a θ' given as follows:

$$\theta' = \mathrm{Ad}\,\phi(\theta) - (d\phi)\phi^{-1} \quad . \tag{1.2}$$

We see that a singularity in θ on X_0 <u>might</u> be cancelled off by $d\phi\phi^{-1}$, leaving θ' singularity-free, enabling one to use θ' to define a connection on $U \times Y$, where U is an open neighborhood of X_0. This enables one to "glue" together

$$U \times Y \quad \text{and} \quad (X-X_0) \times Y \quad ,$$

using (1.1), to obtain a fiber space over all of X, with a non-singular connection. However, this fiber space will typically <u>not</u> be a product. The relevant measure of its "twistedness" will then be a useful "invariant" of the original connection, i.e., "gauge field" θ.

The example to be treated in this chapter should illustrate this process very neatly.

2. SL(2,R) CONNECTIONS WITH ONE-DIMENSIONAL BASE SPACE

Let X be a one-dimensional real space, with coordinate x. (Later on it will be useful to allow x to have complex values.) Let Y be a two-dimensional space with linear coordinate y. Consider the product fiber space

$$X \times Y \to X \quad ,$$

with Y as fiber.

Consider a linear SL(2,R)-connection for this fiber space, defined by a connection form of the following type:

$$\theta = dy - \underline{a}(x)y \tag{2.1}$$

Here,

$$y = \begin{pmatrix} y_1 \\ y_2 \end{pmatrix}$$

is a two-vector,

$$\underline{a} = \begin{pmatrix} a_{11} & a_{12} \\ a_{21} & a_{22} \end{pmatrix}$$

is a matrix.

The solution of $\theta = 0$ are then of the form:

$$y = \underline{y}(x)(y_0) \quad ,$$

where

$$\underline{y}(x) = \begin{pmatrix} y_{11} & y_{12} \\ y_{21} & y_{22} \end{pmatrix}$$

is a 2×2 matrix which satisfies:

$$\frac{d\underline{y}}{dx} = \underline{a}\,\underline{y} \tag{2.2}$$

Suppose now that the point $x = 0$ is a singularity of the matrix $\underline{a}(x)$. It is said to be a Fuchsian or regular singularity if $\underline{a}(x)$ is of the following form:

$$\underline{a}(x) = x^{-1}\underline{b}(x) \quad , \tag{2.3}$$

where $\underline{b}(x)$ is a 2×2 matrix of functions which are continuous in the neighborhood of $x = 0$. Let us suppose, for simplicity, that $\underline{b}(x)$ is analytic in the neighborhood of $x = 0$. Then, the standard Fuchs theorem asserts that $\underline{y}(x)$ can be chosen of the form

$$\underline{y}(x) = x^{A}\underline{z}(x) \quad , \tag{2.4}$$

where A is a 2×2 constant matrix, $\underline{z}(x)$ is analytic and det $\underline{z}(x) \neq 0$ in the neighborhood of $x = 0$. Thus,

$$\frac{d\underline{y}}{dx} = Ax^{A-1}\underline{z}(x) + x^{A}\frac{d\underline{z}}{dx}$$

$$\frac{d\underline{y}}{dx}\underline{y}^{-1} = \left(Ax^{A-1}\underline{z} + x^{A}\frac{d\underline{z}}{dx}\right)(\underline{z}^{-1}x^{-A})$$

$$= Ax^{-1} + x^{A}\left(\frac{d\underline{z}}{dx}\underline{z}^{-1}\right)x^{-A} \tag{2.5}$$

Now set

$$\underline{b}(x) = \frac{d\underline{z}}{dx}\underline{z}^{-1} \tag{2.6}$$

We then have:

$$\underline{a}(x) = Ax^{-1} + x^{A}\underline{b}(x)x^{-A} \tag{2.7}$$

$\underline{b}(x)$ is perfectly well-behaved in the neighborhood $x = 0$. We shall now interpret this fact in bundle language.

3. RESOLUTION OF THE SINGULARITIES OF THE BUNDLE AND/OR CONNECTION AT $x = 0$

Let Y denote R^2, with

$$y = \begin{pmatrix} y_1 \\ y_2 \end{pmatrix} .$$

Consider the product fiber space

$$X \times Y \quad .$$

Let

$$\theta = dy - \underline{a}(x)y \tag{3.1}$$

be the $\underset{\sim}{SL(2,R)}$-valued form that determines a connection. Suppose, however, that $\underline{a}(x)$ (which is 2×2 matrix valued) is of the form (2.3), so that the <u>connection</u> has a singularity at $x = 0$.

Suppose now that (2.7) is satisfied. Construct <u>another</u> product bundle $X \times R^2 \equiv X \times Z$, with fiber now denoted by

$$z = \begin{pmatrix} z_1 \\ z_2 \end{pmatrix} \quad .$$

Set:

$$\theta' = dz - \underline{b}z \tag{3.2}$$

This defines a connection for this bundle which is perfectly smooth at $x = 0$.

Let

$$\phi \colon X \times Y \to X \times Y$$

be defined as follows:

$$\phi(x,z) = (x, x^A z) \quad . \tag{3.3}$$

We see that (modulo the "singularity" at $y = 0$, which we discuss below) ϕ is an "isomorphism" between the <u>bundles</u> and the <u>connections</u>. The original connection (3.1) has been "desingularized".

In order to interpret this better it is now appropriate to "complexify" x. The x-plane can now be "slit" from 0 to ∞, so that $x^A \equiv e^{A \log x}$ becomes well defined. Then ϕ is defined only on

$$(X - X_0) \times Y \quad ,$$

where X_0 is the "slit". "Pasting" these two product structures

$$X \times Y \quad , \qquad X \times Z$$

together "across the slit", as explained in Section 1, defines a _globally defined_ bundle which is _no longer a product_, _but which has a flat connection_. This bundle has "topological invariants", which are, of course, just the classical "monodromy groups". It also has a "flat connection"; the "monodrony group" is then just the holonomy group of the _flat connection_. Thus, we see that we are getting closer to a purely "bundle-theoretic" definition of the "kink invariant" _of solitons_.